Topics in Mining, Metallurgy and Materials Engineering

Series editor

Carlos P. Bergmann, Porto Alegre, Brazil

“Topics in Mining, Metallurgy and Materials Engineering” welcomes manuscripts in these three main focus areas: Extractive Metallurgy/Mineral Technology; Manufacturing Processes, and Materials Science and Technology. Manuscripts should present scientific solutions for technological problems. The three focus areas have a vertically lined multidisciplinarity, starting from mineral assets, their extraction and processing, their transformation into materials useful for the society, and their interaction with the environment.

More information about this series at http://www.springer.com/series/11054

Rasheedat Modupe Mahamood
Esther Titilayo Akinlabi

Functionally Graded Materials

Springer

Rasheedat Modupe Mahamood
Department of Mechanical Engineering Science
University of Johannesburg
Johannesburg
South Africa

and

Department of Mechanical Engineering
University of Ilorin
Ilorin
Nigeria

Esther Titilayo Akinlabi
Department of Mechanical Engineering Science
University of Johannesburg
Johannesburg
South Africa

ISSN 2364-3293 ISSN 2364-3307 (electronic)
Topics in Mining, Metallurgy and Materials Engineering
ISBN 978-3-319-53755-9 ISBN 978-3-319-53756-6 (eBook)
DOI 10.1007/978-3-319-53756-6

Library of Congress Control Number: 2017931574

Printed on acid-free paper

This Springer imprint is published by Springer Nature
The registered company is Springer International Publishing AG
The registered company address is: Gewerbestrasse 11, 6330 Cham, Switzerland

This book is dedicated to God Almighty

Foreword

Material development is the key driver of the world we live in as all areas of human endeavour depend on material development for their performance. The evolution of materials from monolithic material to alloy materials and the development of composite materials is based on the limitation of one class of materials that necessitates the development of other classes of materials. Most applications demand materials with conflicting material properties, which it would be impossible to have in a monolithic material. Furthermore, the alloying of different materials is limited by the thermodynamic behaviour of the constituent materials and the limitation imposed by the degree to which one material can be mixed with other materials. Functionally graded material was born out of the necessity to have two materials combined, and to be able to function and retain their properties after being subjected to harsh working environments. Although, functionally graded material was initially developed for a thermal barrier application, the application of this important advanced material has been increased and used to solve a number of problems in engineering applications, such as for extreme wear resistance and for corrosion resistance applications. Aerospace, automobile, and biomedical applications are some of the areas that are benefitting from this novel material.

The key limitation in using functionally graded material in a number of industries is because of the high cost of production of this material. The area of application of functionally graded material could be increased if the manufacturing process could be simplified and the cost of manufacturing could be reduced. This book was written to shed light on this important material. Various manufacturing processes for this material are also investigated. As an expert in the field of material science and engineering—having published a number of journal articles, and chapters in books, and having presented a large number of conference papers both nationally and internationally—I am in a good position to recommend this book for researchers, industries, professionals and practitioners that are interested in functionally graded materials.

This book provides a number of benefits for researchers in this field, for the professionals in this field, as well as for experts in the field of Mechanical and Material science and Engineering.

The authors of this book are known to me as professional researchers in this field, and they have published widely in this field of research in journals, conference proceedings, book chapters and edited books. Consequently, the authors are in a good position to write this book. The authors began by writing a comprehensive introduction to the book and background of this important material. This should provide the needed information for a new researcher in this field, and it should also provide a better understanding of this material for the readers. The functionally graded materials that exist naturally were also presented. This further brings a better understanding of how nature has been emulated to solve our day-to-day problems.

Different types of functionally graded materials that are produced today and their application areas are presented in this book. This should foster a better understanding on the type of functionally graded material that is needed for specific applications and how it can be produced. The various manufacturing processes that are used to fabricate functionally graded materials are explained in Chap. 3. This chapter is very important, as it provides the readers with the necessary knowledge of these manufacturing processes, as well as their merits and demerits. The use of additive manufacturing technologies for the production of functionally graded materials is the focus of Chap. 4.

This advanced manufacturing technology is the key driver of the next industrial revolution, and the use thereof for making parts with functionally graded materials has been explored in this chapter. The readers will benefit from this advanced technology for the fabrication of parts made with functionally graded materials. A case study presented in Chap. 5 on the use of the laser metal deposition process, an additive manufacturing technology, is of benefit to the reader because the advantage of using additive manufacturing technology for the fabrication of functionally graded materials can be well understood and the advantages appreciated. The book was concluded by the future research need of this interesting research field, while the book was also summarized in this chapter.

The authors have really done justice to this book, and their wealth of experience in this field is well demonstrated.

I strongly recommend this book to you, the reader because of the great benefits it has, and how it could help researchers to further promote development in this research field. The book is composed of six chapters, with each chapter presenting the required knowledge on the subject. The cost of production of functionally graded material is responsible for the prohibitive usage of the material in some application areas; this book provides the benefit of fabricating functionally graded material by using the most cost-effective manufacturing process based on the intended application area. I am writing this foreword as an expert in the field and I want the readers to take the full opportunity of reading this book since it is dedicated to 'functionally graded composite materials', thereby highlighting the importance thereof in a number of human endeavours.

Prof. A.P.I. Popoola
Full Professor
Materials and Metallurgical Engineering
Tshwane University of Technology, Pretoria South Africa

Preface

The introduction of functionally graded composite materials was as a result of the failure of the traditional composite materials when used in hash working environments. The failure of the traditional composite materials occurred along the distinct well-defined interface that exists between the laminate composite materials. The interface causes a high stress concentration at this site, and it promotes crack initiation and eventual crack propagation that result in the ultimate failure of the composite. This process is referred to as 'delamination'. The problem faced with the researchers in Japan during a space plane project (when the functionally graded material was developed) was how this interface can be removed—so that the composite can survive the intended thermal barrier application. The researchers were able to eliminate systematically the sharp interface in the traditional composite material by a gradually changing interface, thereby reducing the stress concentration at this interface, and the functionally graded material developed was able to withstand the extreme working conditions without failing. Subsequently, functionally graded materials have been applied for various engineering applications—apart from the thermal barrier, for which the material was originally developed. Functionally graded materials are advanced composite materials with varying composition, together with the varying properties across the volume of the bulk material.

This book provides the state-of-the art in this advanced composite material. The book was introduced in Chap. 1 by giving a brief historical background of this material. Functionally graded materials occur naturally in nature, as found in human teeth for example, and this was emulated in science, in order to solve engineering problems. Some of these naturally occurring functionally graded materials were also reviewed in Chap. 1. The different types of functionally graded materials that are now produced, together with their areas of application, are reviewed in Chap. 2.

The various conventional manufacturing methods of functionally graded materials for both thin coating and bulk functionally graded materials are analysed, and presented, in Chap. 3. The limitations and many problems that are encountered in the conventional manufacturing process could be overcome by producing functionally graded composite materials—using the additive manufacturing technologies. Additive manufacturing technology is capable of producing three-dimensional components by simply adding materials, layer after layer, as dictated by the computer-aided design model of the part. This process is explored and presented in Chap. 4. Some of the research works in this field are also presented. An experimental analysis of the laser metal deposition process, an additive manufacturing technology, is presented as a case study in Chap. 5 of this book. The summary of the book and the future research direction is presented in Chap. 6.

The organization of the book is as follows:

Chapter 1—An introduction of functionally graded material is presented, together with a brief historical background of functionally graded materials. The presence of functionally graded material in nature is also presented and extensively discussed in this chapter.

Chapter 2—Functionally graded materials are used to solve a number of engineering problems, and it is also used as biomedical implants for the replacement of human tissues. These materials are used to eliminate the stress singularities that occur from the property mismatch between human organs and the implant used to replace them. The different types of functionally graded materials depend on the type of application. The different types of functionally graded materials and their areas of application are presented in this chapter.

Chapter 3—There are different kinds of manufacturing methods for producing functionally graded material (FGM), depending on whether it is functionally graded material thin coating or bulk functionally graded material. The various processing techniques of functionally graded materials, such as physical-vapour deposition, chemical vapour deposition for thin film functionally graded material coatings, and processes, such as the powder metallurgy technique and the centrifugal casting method for bulk functionally graded materials, are analysed in this chapter.

Chapter 4—Additive manufacturing (AM) technology offers many advantages and possibilities for the fabrication of complex three-dimensional products through material addition, rather than the material removal in the conventional machining process. Some of the AM technologies have also been used for the fabrication of complex parts made with functionally graded materials in a single manufacturing run. Some of the AM technologies that are used to produce FGM are the selective laser sintering; the selective laser melting; the laser metal deposition process; and fused deposition modelling. These AM technologies are presented in this chapter, and some of the research work with these technologies for the fabrication of functionally graded materials are also reviewed.

Chapter 5—A case study on the laser metal deposition of functionally graded material (FGM) of titanium alloy composite is presented in this chapter. A functionally graded composite material of an important titanium alloy—Ti64/TiC was fabricated by the Laser Metal Deposition (LMD) process with different TiC percentages of up to 50% on a Ti64 substrate. The microstructures, mechanical and tribological properties of the produced FGM were studied; and the results were compared with the substrate material, as well as with a plain composited material. The microstructure of the fabricated FGM showed a continuous microstructure, without any sharp interface between the substrate and the deposited layers. The results are presented and discussed in detail in this chapter.

Chapter 6—The future research need in functionally graded materials—from the manufacturing process to the material characterization are presented in this chapter, together with an extensive summary of this book.

Johannesburg, South Africa

Rasheedat Modupe Mahamood
Esther Titilayo Akinlabi

Acknowledgements

This work was supported by the University of Johannesburg Research Committee (URC) funds, the L'Oreal-UNESCO For Women in Science, the Department of Higher Education and Training (DHET) South Africa and the National Laser Centre Rental Pool Programme (RPP) contract number NLC - LREHA02-CON-001. We would furthermore like to acknowledge the support of technical reviewers as well as the language and graphic editors who have all contributed to this process. We also value the system of scholarly peer review and the approach that the same adds towards producing research texts in a book such as this that adds value to the body of scientific knowledge.

Contents

Book Description

The necessity of functionally graded materials was borne out of the failure of the traditional composite materials, as a result of their inability to withstand hash working conditions. The failure of the traditional composite materials was as a result of the distinct, or a well-defined interface, that exists between the composite materials. This interface is systematically eliminated in the functionally graded materials, making them to be able to withstand much more extreme working conditions, without failing. Functionally grade materials are advanced composite materials with varying composition and varying properties across the volume of the bulk material. This book provides the much-needed information on the science of composite materials. Various manufacturing methods of producing functionally graded materials are analysed. The shape limitation achievable in the traditional manufacturing process of composite materials can be overcome by producing functionally graded materials using advanced manufacturing processes such as additive manufacturing technologies. Additive manufacturing technology is an advanced manufacturing technology that is capable of producing three-dimensional components by simply adding materials layer after layer, as they are programmed through the computer-aided design model of the part. The specific various additive manufacturing technologies for producing three-dimensional components that are made of functionally graded materials are presented in detail. Experimental analysis of the laser deposition process, an additive manufacturing technology, is also presented in this book. This is the first book that brings to light the capability of additive manufacturing of parts made of functionally graded material, that is not possible with the conventional manufacturing processes. This book also contains an extensive bibliography on this subject.

Chapter 1
Introduction to Functionally Graded Materials

1.1 Introduction

Functionally graded material is an advanced engineering material that is able to survive in a harsh working environment, without losing its properties, and without failing during service. Functionally graded material (FGM) is characterized by a compositional gradient of one material into another, which is totally different from the conventional composite materials, which are either homogeneous mixtures that involve a compromise between the properties of the component materials, or two different materials joined together as in the case of laminate composite materials. Functionally graded materials are materials that are designed to meet varying functionalities [1, 2]. Engineering is constantly turning to nature to seek answers to a number of questions when trying to solve engineering problems. Functionally graded materials are one of such cases. Functionally graded materials exist in materials from nature, such as bones, teeth, wood, and bamboo [1, 3].

Nature has designed these materials because of the service conditions to which these materials are subjected. The high wear-resistant performance that is required from the outside of a human tooth is the reason why nature had to design teeth as a functionally graded material. The outer part of the teeth is made with high wear-resistant material that is referred to as enamel. The inner structure of the teeth is made ductile, because it is needed as a shock absorber, and to improve the fatigue life of the teeth. Human bone is also designed by God in a similar fashion, because of the service requirement expected from the bone.

The idea of functionally graded material for engineering application was first proposed in the early 1980s, when researchers in Japan were confronted with a problem that required a type of composite material that can withstand a very high temperature difference in a space plane project [1]. The application required that one side of the composite materials be subjected to a temperature of about 2000 K, and this temperature should not be transmitted to the other part of the composite material. That is, the body of the plane needs a composite material that will be

R.M. Mahamood and E. Titilayo Akinlabi, *Functionally Graded Materials*,
Topics in Mining, Metallurgy and Materials Engineering,
DOI 10.1007/978-3-319-53756-6_1

exposed to a temperature gradient of approximately 1000 K, between the inside and the outside of the space plane.

The traditional composite materials were unable to withstand such extreme working conditions. The conventional laminate composite materials were tried for this application, but they failed. Each time the traditional composite material failed, the points of failure were usually at the same site, and the modes of failure were always found to be similar. The failure mode was due to delamination, that is, the separation of the laminate composite materials from the site where the two materials were joined together. The point where the failure occurs is usually at this same point at which the two materials were joined together. This site is an area with a high stress concentration factor. This was as a result of the mismatch in the properties of the two materials. The discontinuity at this interface is responsible for the high stress concentration factor that exists at this interface. When the thermal load was applied to this composite material, the thermal mismatch caused the two materials to be separated, because of the different expansion properties of the two materials, which eventually resulted in the failure of the composite material.

The researchers knew that if the sharp interface between the two materials that forms the composite material could be eliminated, then the problem would be solved. The researchers changed this sharp interface into a gradient interface—by gradually introducing the second material into the first material—as against joining 100% of one material and 100% of the second material together. Using this method of gradually introducing the second material into the first material helped to exchange the sharp interface with the gradually changing interface, and the composite material thus developed was able to withstand the intended high temperature application.

This composite material that was developed with the gradient interface is referred to as functionally graded material. The schematic diagram of the conventional laminate composite material and that of the functionally graded material are shown in Fig. 1.1a, b, respectively. The functionally graded material was initially developed for thermal barrier application, but today, FGM has found its application in many areas of human endeavours, such as for high wear-resistant applications such as in the mineral processing industry, or in high penetration resistance applications, such as in bullet-proof application, and also in fire retardant applications.

There are different types of functionally graded materials (FGMs). These include the material gradient in functionally graded material, the microstructural gradient in functionally graded material, and the porosity gradient in functionally graded material. All these are presented in detail in this chapter. The different forms of FGMs and their areas of application are the main focus of Chap. 2. The manufacturing processes of functionally graded materials are discussed in Chap. 3.

Additive manufacturing technologies for the production of FGM are presented in detail in Chap. 4. Experimental analysis of the laser metal deposition process, in a case study, is the focus of Chap. 5, while the book ends with an overall summary, and the future research directions on FGM and are presented in Chap. 6.

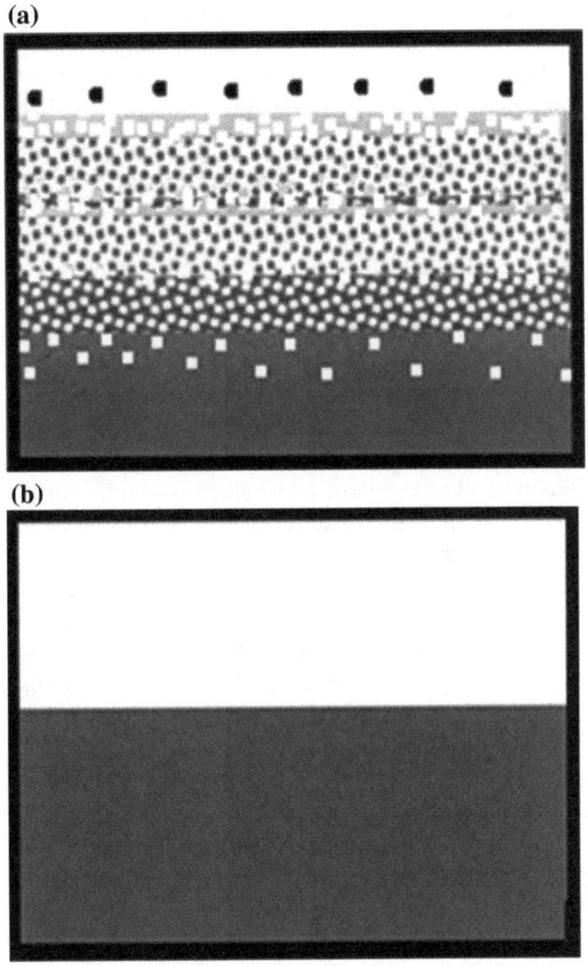

Fig. 1.1 Schematic diagram of **a** functionally graded composite material and **b** traditional laminate composite material

1.2 Brief Background of Functionally Graded Materials

The idea of FGM was started in the early 1980s in Japan, where this material concept was first proposed. The concept of this type of composite material was proposed, in order to reduce the thermal stresses in the conventional laminate composite materials developed for reusable rocket engines [4–6]. Functionally graded materials are characterized by the gradual change in the material composition or structure—with the intention of having a variation in the material properties, along with the change in composition and change in structural direction. Functionally graded materials with graded microstructures can be obtained in monolithic materials by varying the microstructural makeup of the material. This can result in the material possessing some excellent properties that reduce the concentration of thermal stress in such materials.

Functionally graded materials can also be designed in such a way that the material is selectively reinforced in the regions that required to have some special properties. Both the composition and the microstructure that includes the chemical composition, the physical state, and the geometrical configuration can be gradually changed over the entire volume in composite materials. This would result in a corresponding change in the material properties across this volume [5].

The FGM was first conceptualized during the space plane project in Japan for thermal barrier application. The required composite material should be able to withstand a surface temperature of 2000 K, and a temperature gradient of 1000 K, across a less than 10 mm cross section of such composite material. The constant failure of the traditional laminate composite materials being tested at the time of this research was found to occur because of the improper adhesion of the two materials, and the sharp interface that comprised the mismatched properties of the two materials. The problem was solved by replacing this sharp interface with a gradually changing interface that helped to eliminate the site of a high stress concentration factor.

The gradual change in composition of one material to another was what led to the development of the functionally graded material. Although the FGM was initially designed as a thermal barrier material for aerospace structural applications and fusion reactors, the functionally graded materials applications have now been extended to other uses, such as in an environment of extreme wear-resistant application [7, 8].

The metal–ceramic reinforced-based FGMs are able to withstand high-temperature environments by combining the best properties of both materials. The ceramic materials can withstand high temperatures, while the metallic part is able to provide the needed mechanical properties to support the ceramic part. This helps to improve the performance and also to reduce the possibility of delamination. The gradual change in the content of one material into the other material renders the matrix material to see the reinforced material, as a mere impurity that is evenly distributed.

This does not result in a clear boundary between the matrix and the reinforcement, because of the gradual incrementing of the reinforcement material. Hence, the sharp interface is eliminated, and the thermal stress concentrations are thereby greatly reduced. The concept of the FGM has led to the development of material designs with special characteristics, such as an improved Young's modulus, improved tensile strength, and improved wear resistance [1].

Functionally graded materials, with their set of unique mechanical properties offer a number of advantages for the materials that are exposed to an extreme working environment, when compared with the conventional composite materials. The functionally graded material is an important field of research, and it has attracted researchers' attention in the past few decades—with the application area of this novel material constantly expanding. The continuous microstructural control in a material to change the material properties, as well as the material functionality, is also being constantly pursued in the research community. The idea from nature of the natural functionally graded materials have constantly been the driving factor for the development of this artificial functionally graded materials. Wood is a natural FGM from nature that is made up of cellulose in a matrix of lignin [7].

Bone and teeth are other forms of functionally graded materials from nature [3]. These materials were designed by nature to meet their expected service requirements. These great ideas from nature were emulated, in order to design materials that are used to solve engineering problems. The unique characteristics of the functionally graded materials are the ability to tailor a material's properties for a specialized application. These tailored materials possess a number of advantages that render them appropriate for intended and potential applications [9–12]. An appropriately designed FGM at the microstructural level can lead to improved properties in the material [13].

A controlled gradient in the elastic properties of a material helps to improve the failure resistance to wear, and it improves the fracture toughness of such materials [14]. functionally graded materials have appeared prior to the work of the Japanese researchers in 1972, where the graded polymer materials were explored [15, 16]. The first systematic description of FGM appeared in 1995 from the research work reported by Koizumi and Niino [17]. The functionally graded materials that occur in nature are discussed in the next section.

1.3 Functionally Graded Materials in Nature

Nature is always giving the scientist a way, whenever they are confronted with technological problems. This is because nature has a way of designing its work to perform as required, and an example of this is the neurons in the human brain. Scientists have copied this concept to solve a number of world problems through the use of artificial neural networks. Functionally graded material is not an exception, since most of the materials occurring naturally are based on FGMs, where nature has produced them based on functionality that is required from such materials, as well as the working environment in which the material will be used. Most materials produced by nature are composites, and some of these composite materials are functionally graded composite materials. Plant and tree stems, animal bones, and teeth are some of the examples of functionally graded composite materials produced by nature [18–20].

These natural functionally graded materials are designed and optimized for the loading conditions and the working environments to which they are subjected. These materials must be able to perform a variety of functions, and they must be able to perform for a longer period of time also. Bamboo is an example of nature's functionally graded materials that possess continuously graded properties and that are characterized by the spatially varying microstructures produced by non-uniform distributions of the constituent ingredients.

The advantages of the smooth variation in the properties of this material include the reduction of stress concentration and an increased bonding strength [21, 22]. Wood is another functionally graded composite material from nature. It consists of cellulose in the matrix of lignin [23]. Other FGMs from nature include the mollusk shells that have hierarchical architecture-graded ligaments, the exoskeleton of arthropods which has a mineralized fibrous chitin-based nano-composite with hierarchical organization, the spider fang that looks like an injection needle, and the narwhal tooth that has a graded cementum–dentin junction.

Most of the tissues and organs in the human body are made up of naturally occurring functionally graded materials. Examples include the human skin that has a complex multi-layered structural system that consists of the epidermis, the dermis, and the hypodermis. Each layer has different properties, and it also performs different functions; the transitions in human joints are made of functionally graded tissues, such as tendon-to-bone and cartilage-to-bone that contains the ligaments that connects these tissues together, and the human inter-vertebral disc that consists of the annulus fibrosus, the nucleus pulposus, and the cartilage endplates [24].

The bone tissue contains the cancellous spongy bone with changes in pore density and distribution, as shown in Fig. 1.2.

The human teeth with a surface is made up of hard enamel with prismatic hydroxyapatite (HA) crystallites, while the internal core is made up of dentine; a composite system with collagen fibrils and HA is another example. The surface of teeth is made of hard, brittle, and wear-resistant material, while the internal part is made of a soft and flexible material that helps to absorb shock and helps to prolong the life of the hard outer part. The transition from the enamel to dentine is provided

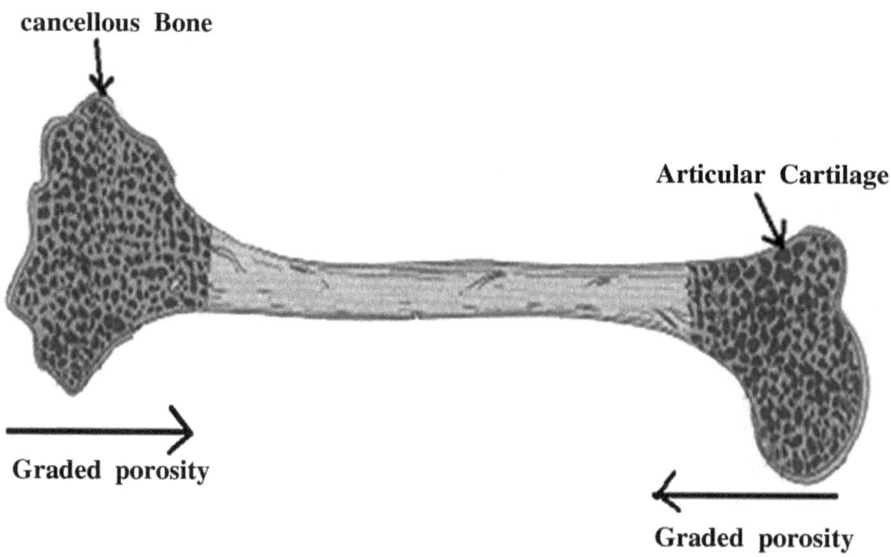

Fig. 1.2 Schematic diagram of human bone showing FGM structures

by an intermediate FGM layers, where the composition gradually changes from one material to the other. This is the reason why there is such a high research interest in the making of most medical implants as functionally graded materials, because it was discovered that the best grafts are often those that are as close to the natural tissue as possible. Readers can consult Oxman [25], for further reading about the functionally graded materials in nature.

Functionally graded materials have been extensively explored for biomedical applications, such as in dental restorations and orthopaedic implants and other medical devices. The lesson from nature has been applied largely to solve many engineering problems. Functionally graded coating and functionally graded bulk materials are explored in the research community, in order to provide the required extreme working environment properties aimed at increasing the service life of engineering parts. There are different types of FGMs that are produced today. The different types of functionally graded materials and their areas of application are the subject of the next chapter.

1.4 Summary

The introduction of functionally graded composite materials, together with a brief historical background of the FGM, has been presented in this chapter. A number of materials occurring naturally are made of functionally graded material, and these form the bases for the use of functionally graded materials in our current day-to-day life. Various materials that were designed to perform different functions that have been designed as functionally graded materials by nature have been reviewed in this chapter.

Acknowledgements This work is supported by the University of Johannesburg Research Commitee fund, the Department of Higher Education and Training (DHET) South Africa, the National Laser Centre Rental Pool Programme (RPP) contract number NLC - LREHA02-CON-001 and L'Oreal-UNESCO For Women in Science.

References

1. Niino, M., Hirai, T., Watanabe, R.: The functionally gradient materials. J. Jpn. Soc. Compos. Mater. **13**, 257–264 (1987)
2. Report on "Fundamental study on relaxation of thermal stress for high-temperature material by tailoring the graded structure". Department of Science and Technology Agency (1992)
3. Knoppers, J.W., Gunnink, J., den Hout, Van, Van Vliet, W.: The reality of functionally graded material products, pp. 38–43. TNO Science and Industry, The Netherlands (2003)
4. Shanmugavel, P., Bhaskar, G.B., Chandrasekaran, M., Mani, P.S., Srinivasan, S.P.: An overview of fracture analysis in functionally graded materials. Eur. J. Sci. Res. **68**(3), 412–439 (2012)

5. Atai, A.A., Nikranjbar, A., Kasiri, R.: Buckling and post-buckling Behaviour of semicircular functionally graded material arches: a theoretical study. In: Proceedings of the Institution of Mechanical Engineers, Part C: Journal of Mechanical Engineering Science, vol. 226, pp. 607–614 (2012)
6. Koizumi, M., Niino, M.: Overview of FGM research in Japan. MRS Bull. **20**(1995), 19–24 (1995)
7. Hon, D., Shiraishi, W.: Cellulose Chemistry, 2nd ed. Marcel Dekker, New York (2001)
8. Wang, S.S.: Fracture mechanics for delamination problems in composite materials. J. Compos. Mater. **17**(3), 210–223 (1983)
9. Shanmugavel, P., Bhaskar, G.B., Chandrasekaran, M., Mani P.S., Srinivasan, S.P.: An overview of fracture analysis in functionally graded materials, Eur. J. Sci. Res. **68**(3), 412–439 (2012)
10. Mahamood, R.M., Akinlabi, E.T., Shukla M., Pityana, S. Functionally graded material: an overview. In: Proceedings of the World Congress on Engineering WCE 2012, vol. 3, pp. 1593–1597 (2012)
11. Jha, D.K., Kant, T., Singh, R.K.: A critical review of recent research on functionally graded plates. Compos. Struct. **96**, 833–849 (2013)
12. Miyamoto, Y., Kaysser, W.A., Rabin, B.H., Kawasaki, A., Ford, R.G.: Functionally Graded Materials: Design, Processing and Applications. Kluwer Academic Publishers, Dordrecht/London/Boston (1999)
13. Wu, X.L., Jiang, P., Chen, L., Zhang, J.F., Yuan, F.P., Zhu, Y.T.: Synergetic strengthening by gradient structure. Mater. Res. Lett. **2**, 185–191 (2014)
14. Suresh, S.: Graded materials for resistance to contact deformation and damage. Science **292**, 2447–2451 (2001)
15. Shen, M., Bever, M.B.: Gradients in polymeric materials. J. Mater. Sci. **7**, 741–746 (1972)
16. Bever, M.B., Duwez, P.E.: Gradients in composite materials. Mater. Sci. Eng. **10**, 1–8 (1972)
17. Koizumi, M., Niino, M.: Overview of FGM research in Japan. MRS Bull. **1**, 19–21 (1995)
18. Amada, S., Munekata, T., Nagase, Y., Ichikawa, Y., Kirigai, A., Yang, Z.: The mechanical structures of bamboos in viewpoint of functionally gradient and composite materials. J. Compos. Mater. **30**(7), 800–819 (1996)
19. Ray, A.K., Das, S.K., Mondal, S., Ramachandrarao, P.: Microstructural characterization of bamboo. J. Mater. Sci. **39**(3), 1055–1060 (2004)
20. Amada, S., Ichikawa, Y., Munekata, T., Nagase, Y., Shimizu, H.: Fiber texture and mechanical-graded structure of bamboo. Composite. Part B **28**, 13–20 (1997)
21. Janssen, J.J.A.: Mechanical Properties of Bamboo. Kluwer Academic Publishers (1991)
22. Paulino, G.H., Jin Z.-H., Dodds Jr., R.H.: Failure of functionally graded materials. In: Karihaloo, B., Knauss, W.G. (eds.) Comprehensive Structural Integrity. Elsevier Science, New York, vol. 2, pp. 607–644 (2003)
23. Hon, D.N., Shiraishi, N.: Wood and Cellulose Chemistry, 2nd ed. Marcel Dekker, New York (2001)
24. Bartel, D.L., Davy, D.T., Keaveny, T.M.: Orthopaedic Biomechanics: Mechanics and Design in Musculoskeletal Systems. Pearson Education, Inc, Upper Saddle River, New Jersey (2006)
25. Oxman, N.: Structuring materiality: design fabrication of heterogeneous materials. Archit. Des. **80**, 78–85 (2010)

Chapter 2
Types of Functionally Graded Materials and Their Areas of Application

2.1 Introduction

Materials with changing composition, microstructure, or porosity across the volume of the material are referred to as the functionally graded material (FGM) [1]. Functionally graded materials (FGMs) are designed with changing properties over the volume of the bulk material, with the aim of performing a set of specified functions [2]. The properties of material in FGMs are not uniform across the entire material, and the properties depend on the spatial position of the material in the bulk structure of the material. Functionally graded materials are designed with varying properties that include changing chemical properties, changing mechanical, magnetic, thermal, and electrical properties. There are FGMs that are designed as stepwise-graded structures, and some are designed to be continuous-graded structures, depending on the areas of application [3–6].

There are different types of areas, in which FGMs are now being used that are different from the initial area of application, for which the material was invented [7].

In this chapter, the different types of FGMs and their areas of application are presented. The different types of FGMs include porosity and pore size gradient-structured FGMs, chemical gradient-structured FGMs, and microstructural gradient-structured FGMs. These different types of functionally graded materials are presented in the next sections.

2.2 Types of Functionally Graded Materials

At the inception of the development of the Functionally graded materials, the concept was to remove the sharp interface that existed in the traditional composite material, and to replace it with the gradually changing interface, which was translated into the changing chemical composition of this composite at this interface

R.M. Mahamood and E. Titilayo Akinlabi, *Functionally Graded Materials*,
Topics in Mining, Metallurgy and Materials Engineering,
DOI 10.1007/978-3-319-53756-6_2

region. The growing interest in this type of material has resulted in different types of FGMs being developed. The type of intended application usually determines the type of FGM to be used. In the biomedical application, for instance, some implants need to really mimic the human organ that they intend to replace or repair, for them to be able to function properly without destroying the surrounding tissues. They also need to be able to last longer in service. In Chap. 1, it was seen that most of the human body is made up of FGMs, because of the functionality requirement. This is one of the reasons why the implants should also be made of FGMs, in order to match the part being replaced or being repaired. The different types of FGMs that are being produced now include the chemical composition gradient FGM, the porosity gradient FGM, and the microstructural gradient FGM. Each of these types of FGMs is discussed in detail in the following sections.

2.2.1 Chemical Composition Gradient Functionally Gradient Materials

This is the type of Functionally graded materials, where the chemical composition is gradually varied, according to the spatial position in the material. This could be in the form of a single phase, or in a multiphased material. A single-phase FGM is produced when the composite is produced from a single phase, as a result of the solubility of the chemical elements of one phase in the other phase. This usually occurs during the sintering process [8]. The gradual change in the distribution of the chemical elements in the single phase results in the formation of the Functionally graded material. According to the phase diagram and thermodynamic limitations, when some materials are added to another material, the material that was added to the other material would be soluble in that material over a range of composition and mixing conditions. Such material would become what is called a single-phase material—but with a varying chemical composition—because of the solubility.

This type of FGM is less common. The most commonly designed and most commonly used Functionally graded materials are the ones with a multiphase chemical composition [9, 10]. The phases and chemical composition are made to vary across the bulk volume of the material. As the composition of material is varied from one material into the other, it will result in different phases with different chemical compositions that would help to achieve the intended application, for which the FGM has been designed. The different phases that are produced are dependent on the compositional quantity of the reinforcing material and the manufacturing conditions—such as the cooling rate and the heat treatment conducted on such material.

In powder metallurgy, the method of producing FGM is by putting the required powder composition layer-by-layer, and this is then followed by powder compaction and thereafter sintering. During the sintering process, some of metallic powders will react to form different chemical compounds and phases. These would vary, according to the spatial position in the Functionally graded material.

2.2.2 Porosity Gradient Functionally Gradient Materials

The porosity gradient functionally graded material is another type of FGM, in which the porosity in the material is made to change with the change in the spatial position in the bulk material. The shape and size of the pore are designed and varied, according to the required properties of the Functionally graded material. The schematic diagram of a typical porosity gradient functionally graded material is shown in Fig. 2.1.

This type of functionally graded material is very important for the biomedical applications, because the natural material they intend to replace consists of functionally graded porosity, and the graded porosity would also help in the integration of the implant and the surrounding tissues. The porosity is important for the healing process of this implant, and it also helps in the blood circulation to the integrated tissues. The graded porosity also helps to reduce the overall weight of the implant and to improve the modulus of elasticity of the implant material to match that of the human tissue. The graded porosity helps to reduce the density of the bio-implant. This is necessary to prevent stress shielding that occurs when the modulus of elasticity of the implant is greater than that of the human bone. Porosity gradient materials are produced by the deposition of powder with a varying mixture of different particle shapes and sizes that would help to produce the needed varying porosity with the changing pore shapes and sizes.

Porosity gradient materials could be porosity density gradation or pore size gradation. The porosity density is produced with the density of porosity changing with respect to the spatial position across the volume of the material. The pore size

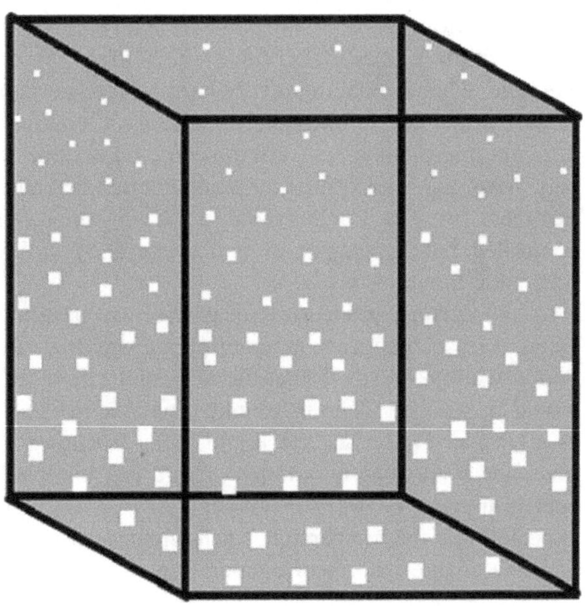

Fig. 2.1 Schematic diagram of porosity-graded FGM

gradient of the FGM, on the other hand, is produced by varying the pore sizes or the pore shape, or both.

The pore size gradation can be achieved by varying the powder particle sizes that are used at different locations in the bulk material during the gradation process. It can also be produced by varying the production processing parameters, or through the use of different sintering parameters to produce the required porosity gradient [11, 12].

The function of pore size gradation is seen in bone implants, where the larger pore sizes in the porosity functionally graded implants are to be implanted in the bone, in order to aid the bone ingrowth, while the smaller pores are useful for the cartilage growth [12]. The function of porosity-graded FGM includes the gradual change in the pore distribution in a porosity-graded FGM that helps in absorbing the shock from one face to the other. It also helps to provide thermal insulation; it helps to aid the catalytic efficiency; and it also helps to relax the electrical and the thermal stresses.

The porosity gradient in a FGM also has an effect on the tensile strength and the Young's modulus of the material. A number of porosity gradient FGMs have been reported in the literature for biomedical application [12–33].

2.2.3 *Microstructure Gradient Functionally Gradient Materials*

Microstructural gradient functionally graded material is another type of FGM, where the microstructure is tailored so that different microstructures are produced in the material, which is made to change gradually, so as to achieve the required properties from the material. Microstructural gradation can be achieved during the solidification process, such that the surface of the material is quenched, for example, when producing a very hard surface property of the material. The core of the same material is allowed to cool down slowly, which would help to produce different microstructures from those on the surface of the material to the innermost part. Also, the microstructural gradation can be achieved through a controlled heat treatment process. For example, a varying microstructure can be produced by a controlled heat treatment of a titanium-alloy cylindrical part, as shown by the schematic diagram in Fig. 2.2.

A functionally graded microstructure can be achieved by first allowing a liquid metal, whose melting temperature is lower than the melting temperature of the titanium alloy, and such that the recrystallization temperature of the titanium alloy would be reached when this molten metal is run in a kind of heat exchanger setup, as shown in Fig. 2.2a. The liquid metal is allowed to run for a certain period of time, and subsequently withdrawn, and then the part is allowed to cool down. The heat is transferred from the inner part of the cylinder to the external part. It is expected that the temperature of the innermost part of the cylinder is much higher than that of the outermost temperature.

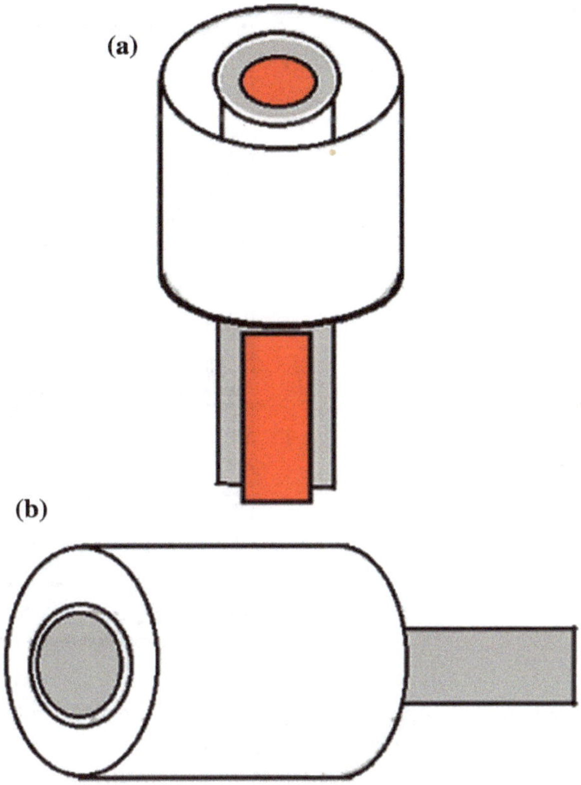

Fig. 2.2 Schematic diagram of cylindrical part subjected to a flow of **a** liquid metal and **b** cooling water

This would result in a varying microstructure, as the part is allowed to cool down. The outer part would behave like a heat sink, and the microstructure of this area would be larger because of the grain growth, while the innermost microstructure would be smaller and equiaxed, because of the refined microstructure during the recrystallization process and the slow cooling rate. In another type of microstructural gradation process, the cylinder could be heated to a certain temperature, and then cooling water could be run through the inner part of the cylinder—also in a heat exchanger setup, as shown in Fig. 2.2b. The inner part of the cylinder would be subjected to rapid cooling, thereby causing the formation of a non-equilibrium microstructure at this face. The innermost microstructure would consist of a martensitic microstructure that is harder, while the microstructure of the cylinder far away from the innermost part would be completely different from the microstructure on the outer part of the cylinder.

A schematic diagram of a typically graded microstructure is shown in Fig. 2.3. This is because, the outermost part of the cylinder would cool down more slowly, and this would favour the formation of a more equilibrium microstructure, and a largely equiaxed microstructure would be produced. The gradients in the microstructure due to the heat treatment could also cause some changes in the

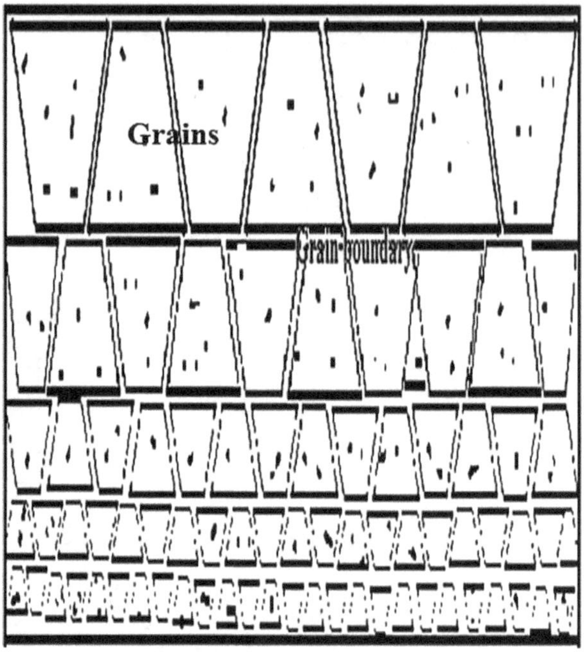

Fig. 2.3 Schematic diagram of graded microstructure

elemental composition, and in some cases, the intermetallic phase could be produced in the graded microstructure.

The graded microstructure would result in a gradual change of the material properties with respect to position, since the microstructure is dependent on the position in the FGM, and because the microstructure is directly related to the properties of the material.

The microstructural gradient FGMs find their application in components that must have a very hard surface to resist wear, and a tough core to resist the high impact that occurs during the operation. An example of this type of Functionally graded material includes case-hardened steel, cams or ring gear, bearings or shafts, and turbine applications [34, 35]. The areas of application of the different types of FGMs are presented in the next section.

2.3 Areas of Application of Functionally Graded Materials

The important characteristics of the FGM have made them to be favoured in almost all the human areas of endeavour. Functionally graded materials are currently being applied in a number of industries, with a huge potential to be used in other applications in the future. The current applications and futuristic application of the FGM are presented in this section. The current areas of applications include

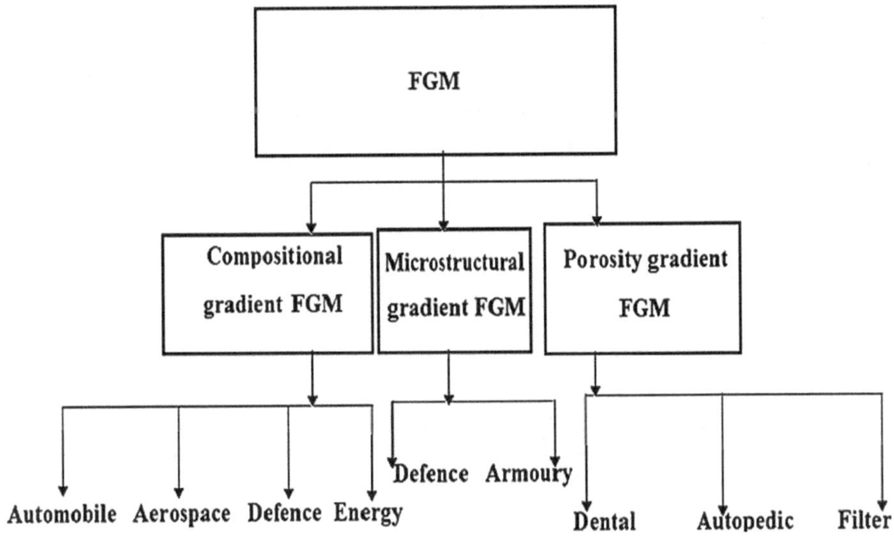

Fig. 2.4 Areas of applications for the three types of functionally graded materials

aerospace, automobile, biomedical, defence, electrical/electronic, energy, marine, opto-electronics, and thermoelectronics. Figure 2.4 shows the different types of FGMs and their application areas. FGM offers great promise in applications with harsh operating conditions, or example, for wear-resistant linings for handling large heavy abrasive ore particles in the mining industry, for the rocket heat shields, for the heat engine components, for heat exchanger tubes, for the plasma facings for fusion reactors in nuclear reactor plant, for thermo-electric generators, and in the electrical insulating applications.

Functionally graded materials are also ideal for reducing the mismatch in the thermo-mechanical properties in metal–ceramic bonding that help to prevent debonding. The future demands for functionally graded materials are in such applications, where extraordinary mechanical, thermal, and chemical properties are required, and which must be able to sustain severe working environments. These potential future application areas include applications, where the structural and the engineering uses require a combination of incompatible functions, such as hardness and toughness.

The future application areas of functionally graded materials will also expand, when the production costs of these important engineering materials are reduced. The fabrication processes of functionally graded materials are explained in detail in Chap. 3 and the use of additive manufacturing technologies for the production of FGMs is presented in Chap. 4.

Although some applications are more interested in the reliability of the FGM, rather than the cost of such materials, such niche industries include aerospace and nuclear energy. On the other hand, the cost of production of functionally graded materials is important in some applications such as the cutting tools, machine parts, and engine components. The use of functionally graded materials is now seen as one of the most important, effective, and efficient materials for promoting sustainable development in industries. Some of these applications, such as aerospace, automobile, biomedical, defence, energy, and the marine industry, are presented and explained in the following subsections.

2.3.1 The Aerospace Industry

The initial application where functionally graded materials were developed was for space plane bodies. The application of this novel material is increased over the years in the aerospace industry. Most aerospace equipment and structures are now made of functionally graded materials. These include the rocket engine components, the spacecraft truss structure, the heat exchange panels, and some structures, such as the reflectors, the solar panels, the camera housing, the turbine wheels, the turbine blade coatings, the nose caps, the leading edge of missiles, and space shuttles. Functionally graded materials are also used for the structural walls that combine thermal and sound insulation properties. Automobiles are another industry, in which functionally graded materials have been used. These will be presented in the next subsection.

2.3.2 The Automobile Industry

The use of functionally graded materials in the automobile industry is still limited at the moment, because of the high cost of production of functionally graded materials. However, the material is being used in very important parts of the automobile, where the present high cost justifies its use. The present applications include the engine cylinder liners for diesel engine pistons, for the leaf springs, for the spark plugs, the combustion chambers, the drive shafts, the shock absorbers, the flywheels, some car body parts, the window glass, and racing car brakes. Also, functionally graded materials are used in enhanced body coatings for cars, and that includes the graded coatings with particles, such as dioxide/mica.

2.3.3 Biomedical

The human body is made up of a number of functionally graded materials, which includes the bones and the teeth. These are the most replaced human body parts, as a result of damage to these parts, or as a result of the natural ageing process. The engineering materials that are biocompatible are used for their replacements. The natural parts that these materials replace are functionally graded materials in nature. This is why the majority of functionally graded materials used in the biomedical industry are used for implants. The porosity gradient functionally graded materials are most commonly used in this industry, because their properties are very close to those of the parts they intend to replace. Examples of where the porosity gradient FGM is used in the biomedical industry include the following: In the permanent skeletal replacement implants, graded porosity helps to minimize the stress shielding [13]. The gradient porous titanium dental implants also help to improve the osseo-integration properties of the implant [14]. The graded porous hydroxyapatite (HA) mimics the bimodal structure of the human bone (cortical and cancellous), which helps to promote the new tissue growth, and also with the desired mechanical properties [17–19].

2.3.4 Defence

The ability of the FGM to offer penetration-resistant properties by inhibiting crack propagation is an attractive property that makes the material favoured in the defence industry. The functionally graded materials are used in the defence industry in applications, such as bullet-proof vests, the traditional Japanese sword, and in armour plates. Another key area of application of functionally graded materials is in the body of bullet-proof vehicles.

2.3.5 Energy

The energy industries are constantly in need of different types of functionally graded materials, in order to improve the efficiency of some of their equipment. Some of the applications of the functionally graded materials in the energy industry include the inner wall of nuclear reactors, the thermo-electric converter for energy conversion, the solar panel, the solar cells, the tubes and pressure vessels, the graded electrode for the production of solid oxide fuel, the piezo-electric functionally graded materials for the ultrasonic transducer, the dielectric, the fuel cell, the turbine blade coatings, and for thermal barrier coatings.

2.3.6 Electrical/Electronics

Functionally graded materials are used in the electrical and the electronic industries in a number of ways. These include in the relaxation of the field stress in the electrode and the field–spacer interface [36, 37], in the diodes, in the semiconductors, for insulators, and for the production of sensors. The thermal-shielding elements in the micro-electronics are also made from the carbon nanotube functionally graded materials.

2.3.7 *Marine*

Functionally graded materials also find their application in the marine industry. The applications of functionally graded materials in the marine and sub-marine industry include in the propeller shaft, the diving cylinders, the sonar domes, the composite piping system, and in the cylindrical pressure hull.

2.3.8 *Opto-Electronics*

Functionally graded materials find their application in the opto-electronic industry for the production of such parts as those that are made with the optical fibre materials, the lens, the GRINSH lasers, the highly efficient photo dectectors, the solar cells, the tunable photodetector, the magnetic storage media, and in the production of semiconductors—with a varying refractive index.

2.3.9 *Sport*

Functionally graded materials are used in a number of sporting equipment, such as the golf clubs, tennis rackets, and skis. These are all made of functionally graded materials.

2.3.10 *Others*

The application of functionally graded materials also includes, but is not limited to, the cutting tools and dies to improve the thermal strength of the cutting tool and die, razor blades of iron-aluminide/stainless steel [38], in the safety equipment, such as Firefighting air bottles, the fire-retardant doors, the eyeglass frames, and the helmets. Others include the MRI scanner cryogenic tubes, the pressure vessels, the fuel tanks, the laptop cases, the musical instruments, and the X-ray tables. The applications of FGMs in Japan have also been presented by Miyamoto [39]. Readers can consult the material for further readings. The area of application of functionally graded materials is expected to increase, if the cost of production of this material is reduced in the future.

2.4 Summary

The different types of functionally graded materials and their application areas have been presented in this chapter. The functionally graded materials that were initially developed were chemical compositional gradient functionally graded materials, and it was developed for thermal bearer application. The high interest in the functionally graded materials in the research community has made this novel material to evolve into different types, while the areas of application have also expanded greatly. The three main types of functionally graded materials presented in this chapter are the chemical composition gradient functionally graded materials, the microstructure gradient functionally graded materials, and the porosity gradient functionally graded materials. The porosity gradient functionally graded materials are widely used in the biomedical application as medical implants, because they are designed to mimic the human organs, which are functionally graded materials in nature.

Different areas of application of this type of functionally graded material and other types of functionally graded materials have also been presented in this chapter. The areas of application of functionally graded materials are also expected to increase, if the cost of producing functionally graded materials is reduced.

Acknowledgments This work is supported by the University of Johannesburg Research Council, the Department of Higher Education and Training (DHET) South Africa, the National Laser Centre Rental Pool Programme (RPP) contract number NLC-LREHA02-CON-001 and L'Oreal-UNESCO For Women in Science.

References

1. Niino, M., Hirai, T., Watanabe, R.: The functionally gradient materials. J. Jpn. Soc. Compos. Mater. **13**, 257–264 (1987)
2. Report on. Fundamental study on relaxation of thermal stress for high-temperature material by tailoring the graded structure. In: Department of Science and Technology Agency (1992)
3. Dumont, A.-L., Bonnet, J.-P., Chartier, T., Ferreira, J.M.F.: $MoSi_2/Al_2O_3$ FGM: elaboration by tape casting and SHS. J. Eur. Ceram. Soc. **21**, 2353–2360 (2001)
4. Saiyathibrahim, A., Mohamed, N.S.S., Dhanapal, P.: Processing techniques of functionally graded materials—a review. In: International Conference on Systems, Science, Control, Communication, Engineering and Technology, pp. 98–105 (2015)
5. Nemat-Alla, M.M., Ata, M.H., Bayoumi, M.R., Khair-Eldeen, W.: Powder metallurgical fabrication and microstructural investigations of aluminum/steel functionally graded material. Mater. Sci. Appl. **2**, 1708–1718 (2011)
6. Jin, X., Wu, L., Sun, Y., Guo, L.: Microstructure and mechanical properties of ZrO_2/NiCr functionally graded materials. Mater. Sci. Eng. A **509**, 63–68 (2009)
7. Wosko, M., Paszkiewicz, B., Piasecki, T., Szyszka, A., Paszkiewicz, R., Tlaczala, M.: Applications of functionally graded materials in opto-electronic devices. Optica Applicata **35**(3), 663–667 (2005)
8. Bharti, I., Gupta, N., Gupta, K.M.: Novel applications of functionally graded nano, opto-electronic and thermo-electric materials. Int. J. Mater. Mech. Manuf. **1**, 221–224 (2013)
9. Mahamood, R.M., Akinlabi, E.T.: Laser-metal deposition of functionally graded Ti6Al4V/TiC. Mater. Des. **84**, 402–410 (2015)
10. Mahamood, R.M., Akinlabi, E.T., Shukla M., Pityana, S.: Functionally graded material: An overview. In: Proceedings of the world congress on engineering WCE 2012, vol. 3, pp. 1593–1597 (2012)
11. Mahamood, R.M., Akinlabi, E.T.: Modelling of process parameters influence on degree of porosity in laser-metal deposition process. In: Yang G-C et al. (eds) Transactions on Engineering Technologies. Springer, pp. 31–42 (2015)

12. Mia, X., Sun, D.: Graded/gradient porous biomaterials. Materials **3**, 26–47 (2010). doi:10.3390/ma3010026
13. Thieme, M., Wieters, K.P., Bergner, F., Scharnweber, D., Worch, H., Ndop, J., Kim, T.J., Grill, W.: Titanium-powder sintering for preparation of a porous functionally graded material destined for orthopaedic implants. J. Mater. Sci. **12**, 225–231 (2001)
14. Suk, M.J., Choi, S.I., Kim, J.S., Kim, Y.D., Kwon, Y.S.: Fabrication of a porous material with a porosity gradient by a pulsed electric-current sintering process. Met. Mater. Int. **9**, 599–603 (2003)
15. Woodfield, T.B.F., van Blitterswijk, C.A., de Wijn, J., Sims, T.J., Hollander, A.P., Riesle, J.: Polymer scaffolds fabricated with pore-size gradients as a model for studying the zonal organization within tissue-engineered cartilage constructs. Tissue Eng. **11**, 1297–1311 (2005)
16. Oh, S.H., Park, I.K., Kim, J.M., Lee, J.H.: In vitro and in vivo characteristics of PCL scaffolds with pore-size gradients fabricated by a centrifugation method. Biomaterials **28**, 1664–1671 (2007)
17. Tampieri, A., Celotti, G., Sprio, S., Delcogliano, A., Franzese, S.: Porosity-graded hydroxyapatite ceramics to replace natural bone. Biomaterials **22**, 1365–1370 (2001)
18. Werner, J.P., Linner-Krcmar, B., Friess, W., Greil, P.: Mechanical properties and in vitro cell compatibility of hydroxyapatite ceramics with graded-pore structure. Biomaterials **23**, 4285–4294 (2002)
19. Rodriguez-Lorenzo, L.M., Ferreira, J.M.F.: Development of porous ceramic bodies for applications in tissue engineering and drug-delivery systems. Mater. Res. Bull. **39**, 83–91 (2004)
20. Lu, W.W., Zhao, F., Luk, K.D.K., Yin, Y.J., Cheung, K.M.C., Cheng, G.X., Yao, K.D., Leong, J.C.Y.: Controllable porosity hydroxyapatite ceramics as a spine cage: fabrication and properties evaluation. J. Mater. Sci. **14**, 1039–1046 (2003)
21. Lee, B.T., Kang, I.C., Gain, A.K., Kim, K.H., Song, H.Y.: Fabrication of pore-gradient Al_2O_3–ZrO_2 sintered bodies by fibrous monolithic process. J. Eur. Ceram. Soc. **26**, 3525–3530 (2006)
22. Sherwood, J.K., Riley, S.L., Palazzolo, R., Brown, S.C., Monkhouse, D.C., Coates, M., Griffith, L.G., Landeen, L.K., Ratcliffe, A.: A three-dimensional osteo-chondral composite scaffold for articular cartilage repair. Biomaterials **23**, 4739–4751 (2002)
23. Chen, G., Sato, T., Tanaka, J., Tateishi, T.: Preparation of a biphasic scaffold for osteo-chondral tissue engineering. Mater. Sci. Eng. **26**, 118–123 (2006)
24. Rowe, J.R., Russell, H., Lare, P.J., Hahn, H.: Surgical implants having a graded porous coating. U.S. Patent No. 4542539 (1985)
25. Miao, X., Hu, Y., Liu, J., Tio, B., Cheang, P., Khor, K.A.: Highly interconnected and functionally graded porous bioceramics. Key Eng. Mater. **240–242**, 595–598 (2003)
26. Droschel, M., Hoffmann, M.J., Oberacker, R., Both, H.V., Schaller, W., Yang, Y.Y., Munz, D.: SiC-ceramics with tailored porosity gradients for combustion chambers. Key Eng. Mater. **175–176**, 149–162 (2000)
27. Cichocki Jr., F.R., Trumble, K.P., Rodel, J.: Tailored porosity gradients via colloidal infiltration of compression-moulded sponges. J. Amer. Ceram. Soc. **81**, 1661–1664 (1998)
28. Harley, B.A., Hastings, A.Z., Yannas, I.V., Sannino, A.: Fabricating tubular scaffolds with a radial pore size gradient by a spinning technique. Biomaterials **27**, 866–874 (2006)
29. Bretcanu, O., Samaille, C., Boccaccini, A.R.: Simple methods to fabricate bioglass-derived glass-ceramic scaffolds exhibiting a porosity gradient. J. Mater. Sci. **43**, 4127–4134 (2008)
30. Li, R., Liu, J., Shi, Y., Du, M., Xie, Z.: 316L Stainless steel with gradient porosity fabricated by selective laser melting. J. Mater. Eng. Perform. **19**(5), 666–671 (2010)
31. Muthutantri, A., Huang, J., Edirisinghe, M.: Novel preparation of graded porous structures for medical engineering. J. R. Soc. Interface **5**, 1459–1467 (2008)
32. Macchetta, A., Turner, I.G., Bowen, C.R.: Fabrication of HA/TCP scaffolds with a graded and porous structure using a camphene-based freeze-casting method. Acta Biomater. **5**, 1319–1327 (2009)

33. Hsu, Y.H., Turner, I.G., Miles, A.W.: Fabrication of porous bioceramics with porosity gradients similar to the bimodal structure of cortical and cancellous bone. J. Mater. Sci. Mater. Med. **18**, 2251–2256 (2007)
34. Schneider, M.J.: The Timken Company, and Madhu S. Chatterjee, bodycote introduction to surface hardening of steels. In: Dossett, J., Totten, G.E. (eds) ASM Handbook, Steel Heat-Treating Fundamentals and Processes, vol. 4 (2013)
35. Lu, L., Chekroun, M., Abraham, O., Maupin, V., Villain, G.: Mechanical properties estimation of functionally graded materials using surface waves recorded with a laser interferometer. NDT and E Int. **44**(2), 169–177 (2011)
36. Shumiya, H., Kato, K., Okubo, H.: Feasibility studies on FGMs (functionally graded materials) application for gas insulated equipment. In: IEEE Conference on Electrical Insulation and dielectric Phenomena, pp. 360–363 (2004)
37. Kato, K., Kurimoto, M., Shumiya, H., Adachi, H., Sakuma, S., Okubo, H.: Application of functionally graded material for solid insulator in gaseous-insulation systems. IEEE Trans. Dielectr. Electr. Insul. **13**(2), 362–372 (2006)
38. Miyamoto, Y., Kaysser, W.A., Rabin, B.H., Kawasaki, A., Ford, R.G.: Functionally Graded Materials: Design, Processing and Applications. Kluwer Academic, Boston (1999)
39. Miyamoto, Y.: The applications of functionally graded materials in Japan. Mater. Technol. **11**(6), 230–236 (1996)

Chapter 3
Processing Methods of Functionally Graded Materials

3.1 Introduction

Functionally graded material (FGM) is an advanced composite materials with the sharp interface that exists in the traditionally composite materials being replaced with the gradually changing interface that helps the material to be able to survive in extreme working environments. Functionally graded materials (FGMs) could be in the form of thin coatings that are applied to the surface of a material—with the intention of improving the surface properties of such materials, or it could be in form of bulk material, in which the material properties are changing across the whole volume of the material. The composition, the microstructure, or the porosity could be designed so that the content is varied across the volume of a material, in order to produce what are known as functionally graded materials that are able to perform the functions that they are designed to perform.

There are different types of deposition processes that are used to produce the thin films of functionally graded material coatings. These include the physical vapour deposition process, the chemical vapour deposition process, and the self-propagating high-temperature synthesis (SHS) process, or a combination of some of these processes [1, 2]. The physical vapour deposition process consists of processes, such as the evaporation-based processes, the sputtering-based processes, and the spray-based processes.

Thin-film functionally graded materials are used in some applications that require the surface of the material to have properties that are completely different from those of the bulk material. In some applications, it is required that the material should have bulk functionally graded materials, because of the extreme nature of the working environments the material will be subjected. A number of fabrication technologies have been used to produce the bulk functionally graded materials. These include the powder metallurgy method, the centrifugal casting method, the slip casting method, and the tape casting method [1, 3, 4].

R.M. Mahamood and E. Titilayo Akinlabi, *Functionally Graded Materials*,
Topics in Mining, Metallurgy and Materials Engineering,
DOI 10.1007/978-3-319-53756-6_3

All these processing techniques involve basically two steps, namely the building of the homogeneous-graded structure and the consolidation of the graded material. The use of the physical vapour deposition process and the chemical vapour deposition process for the production of thin functionally graded material coatings are explained in detail in this chapter. Additionally, the use of powder metallurgy and the casting-based processing techniques for the fabrication of the bulk functionally graded materials are explained in this chapter, together with their merits, demerits, and areas of application. An additive manufacturing method, which is an advanced manufacturing process, is also used for the production of thin functionally graded materials as well as bulk functionally graded materials. This will be the focus of the next chapter.

3.2 Physical Vapour Deposition

The physical vapour deposition (PVD) process comprises a vaporisation-based coating method that is used for the production of thin-film coatings and thin functionally graded material coatings. The PVD is achieved through the vaporisation of the material that is to be deposited, and the vapourised material (atomised material) is then transferred onto the surface that needs to be coated [5]. For the functionally graded coatings, two crucibles are used, with each of the crucibles containing the materials to be deposited. The PVD is an alternative process to the electroplating process—but with better properties than the electroplating process.

Usually, the material to be deposited is in a solid form at the beginning of the deposition process, but this solid material is then gradually vapourised from the surface until the solid material gradually diminishes, as it is being used up in the coating process. There are different ways of atomising the solid material to be deposited onto the surface to be coated, depending on the type of PVD process that is being employed. These include the

evaporation-based PVD process, the sputtering-based PVC process, and the plasma spray-based PVD system. All these physical vapour deposition processes take place under vacuum, and they involve three to four basic processes, depending on the type of material that is being deposited.

The evaporation-based PVD is explained in the next subsection.

3.2.1 *Evaporation-Based PVD Process*

The evaporation-based PVD process involves the evaporation of the material to be deposited through melting, and subsequently, vaporising the material using an appropriate heat source. The heat source, which could be electrical resistance, hot filament, electric arc, electron beam, or laser beam, is located very far away from the substrate or the material to be coated. There are other forms of heat sources that are also used to produce the vapour of the target material. These include molecular beam epitaxy, ion beam-assisted evaporation, and the discharge-based evaporation method [6].

The atomised material that is produced from the vapour of the coating material—also known as the target material—is then transported to the substrate, where the coating takes place in the vacuum. The energy from the heat source is used to raise the temperature of the material for the coating—until the material is evaporated or sublimated. The four basic stages that are involved in this process are the evaporation stage, the transportation stage, the reaction stage (optional), and finally, the deposition stage.

During the evaporation stage, the material to be deposited is placed in a crucible, and it is then bombarded by the high energy from the heat source, such as the thermal energy from the resistance heater (thermal evaporation), or the beam of electrons from the electron-based system, which causes the material to be heated up, melted, and then vapourized. The schematic diagram of the thermal and the electron beam evaporation-based PVD process is shown in Fig. 3.1a, b, respectively.

The vapour produced from the surface of this material is then transported into the second stage to the substrate that is located at a distance from the source material. The third stage is the reaction stage, which is optional. It is only needed when the required coating needs to be in the form of a metal oxide, a carbide or a nitride. At this stage, the reaction between the atomised material vapour and the reactive gases, such as oxygen, methane, or nitrogen introduced into the system, takes place during the transportation stage—before reaching the surface of the substrate, or even at the surface of the substrate—depending on the type of process that is used.

If the pure form of the source or the target material is required, then the reaction stage is eliminated. The final stage is the actual deposition process that involves the coating of the atomised material onto the surface of the substrate. For the FGM coatings, the vapourised materials from the two crucibles are deposited simultaneously onto the surface of the substrate material, at a controlled rate, depending on the compositional ratio of the two materials that are required from the material design. The advantages of the evaporation-based PVD process include high deposition rate, simple and easy to use, conductor materials in electronic circuits that can be deposited, and dielectric and optical coatings, which can be easily achieved [7].

Some of the disadvantages of the evaporation-based PVD include: the evaporation coefficient is affected by the contamination of the target material surface; the deposition rate depends on the substrate-to-target material geometry; the coating thickness on the surface of the coated material is distance-dependent, and hence, the coating thickness is not uniform; the difficulty in depositing materials with high melting points; the difficulty in achieving uniform heating; and the reactions between the target material and the heating container. These issues could also be a problem [7].

The sputtering-based PVD is presented in the next subsection.

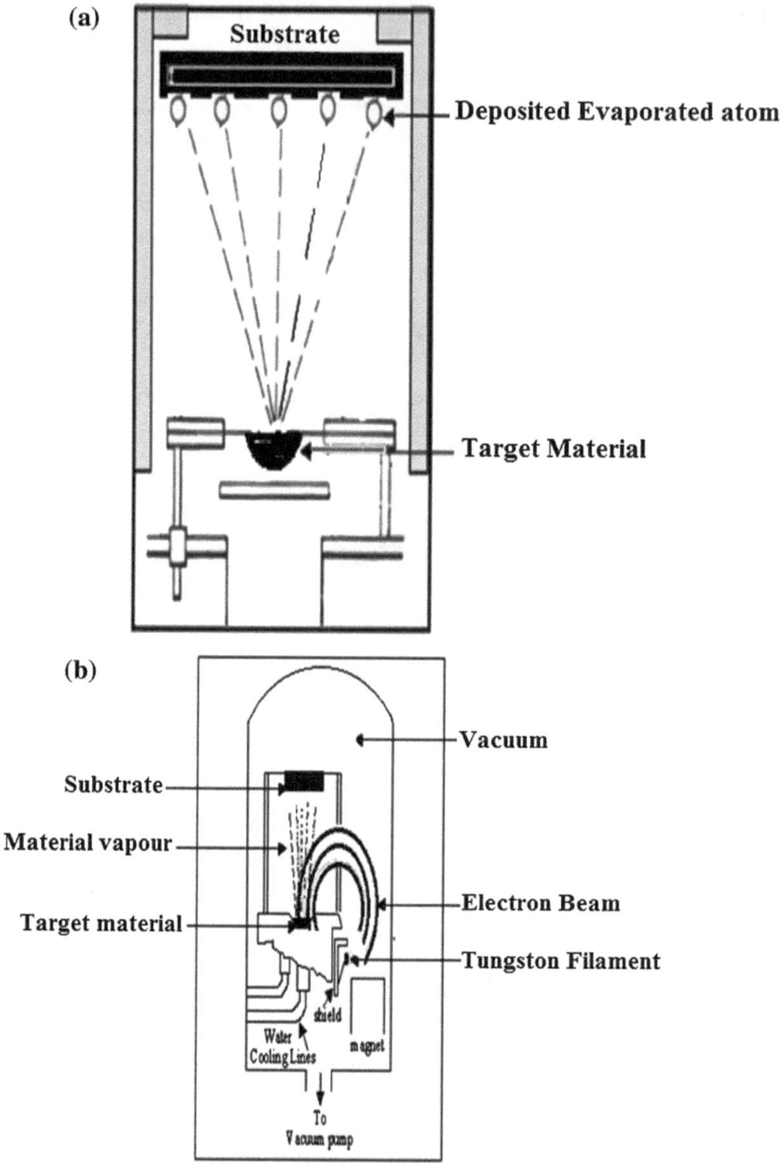

Fig. 3.1 Schematic diagram of **a** thermal evaporation-based PVD process and **b** Electron beam-based evaporation PVD process

3.2.2 Sputtering-Based PVD

There are different types of sputtering-based physical deposition processes, depending on the type of power source used. These include the direct current (DC) or the diode sputtering-based PVD process, the radio frequency (RF) sputtering-based PVD process, the magnetron sputtering-based PVD process, and the reactive sputtering-based PVD process. In the sputtering-based process, the sputtering gas (which is an inert gas) ions, which are under high kinetic energy, are made to constantly collide with the surface of the target material. This results in the ejection of the atoms from the target material. It is these ejected atoms of the target materials that are eventually deposited on to the substrate. These ions of the inert gas are accelerated into the surface of the target material. Some of the surface atoms of the target materials are sputtered, or removed, or ejected from the surface of the target material, and then the sputtered atoms flow across the chamber to be deposited on the substrate.

Four stages are also involved in this process, namely, the sputtering of the target atoms into the gaseous state; the transportation of the sputtered atoms through the plasma medium onto the substrate; the reaction of the sputtered atoms with the reactive gases, which is optional; and the final deposition of the atoms on the substrate. The schematic diagram of sputtering-based PVD is shown in Fig. 3.2. The transportation stage involves the target atoms passing through the sputtering gas, or the inert gas, and the plasma environment, and the collision of the electrons in the plasma environment with the neutral atom of the inert gas, to form ions. More electrons are required to help maintain the plasma formation.

The rate of the target material atoms ejection is used to describe the yield of the sputtering process. The yield is typically in the range of 0.01–4 (atoms/ion). The yield is dependent on the mass of the target material, the available energy of the ions of the sputtering gas, the binding energy of the atoms in the target material, and the incident angle of the ions of the sputtering gas.

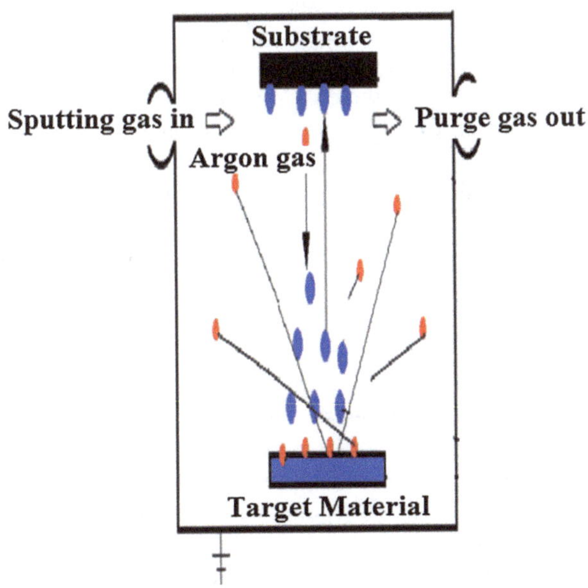

Fig. 3.2 Schematic diagram of sputtering-based PVD process

Some of the advantages of the sputtering-based PVD process include the following: It is not a line-of-sight processes—except that the pressure in the system is reduced and the inelastic collisions add energy to the system.

In the DC sputtering system, the substrate material is the anode, while the target material is the cathode. The puttering yield is dependent on the sputter voltage, which is typically ranging from −2 to −5 kV [7]. The deposition rate is dependent on the inert gas pressure, which is also dependent on the voltage. The higher the voltage, the higher the gas pressure, and the higher the yield, which will result in a high deposition rate.

The optimum deposition rate is about 100 m Torr, which can be achieved with a compromise between the increasing number of the inert gas ions and the increase in the scattering of the inert gas ions and the neutral atoms (plasma is the fourth state of matter, where the ions and the neutrons of an atom are splitted). If the number of the inert gas ions can be increased without increasing the number of the neutrons, then the system can be operated at a lower pressure. The function of the negative bias put on the substrate will help to control the number of neutral atoms.

Because the substrate is constantly being bombarded by the deposited atoms and the plasma, during the deposition process, this causes the formation of the neutral atoms as the by-product. These neutral atoms can contaminate the deposition process; hence, they need to be removed through the negative bias on the substrate. The substrate temperature also needs to be controlled, so as to control the property of the deposited film. The higher the sputter voltage, the greater the increase in the bombardment rate, and thus, increasing the substrate temperature. By increasing the substrate bias, then the substrate temperature can be reduced.

The schematic diagram of the DC sputtering process is shown in Fig. 3.3. The DC sputtering is used with the electrically conductive target materials. The process is easy to control, and it is also cheaper than other types of sputtering methods. The DC sputtering cannot be used for insulating materials, since it requires a very high voltage to sputter such materials. Additionally, in the DC sputtering system, there is constant positive charge building up on the cathode (the target material). This can be avoided using the alternating potential, which is used in the radio frequency sputtering system.

The radio frequency (RF) sputtering is similar to the DC sputtering, but the substrate bias is replaced with the switching potential [14]. The sputtering is alternated between the source material and the substrate materials, as shown in Fig. 3.4.

The RF power can be used to deposit all the materials. It finds most of its applications in the deposition of films from dielectric-sourced materials. The deposition rate is dependent on the relative duty cycle. The deposition rate is lower than the DC sputtering rate. This is because of the difference in the mobility rate of electrons and ions in the plasma environment. Consequently, the electron flux on the substrate is higher, and this may cause a significant increase in temperature of the substrate. The RF sputtering is very expensive because of the high cost of the RF power, and it also operates at a very high frequency—typically of about 13.56 MHz. That is why the RF sputtering deposition process is generally limited

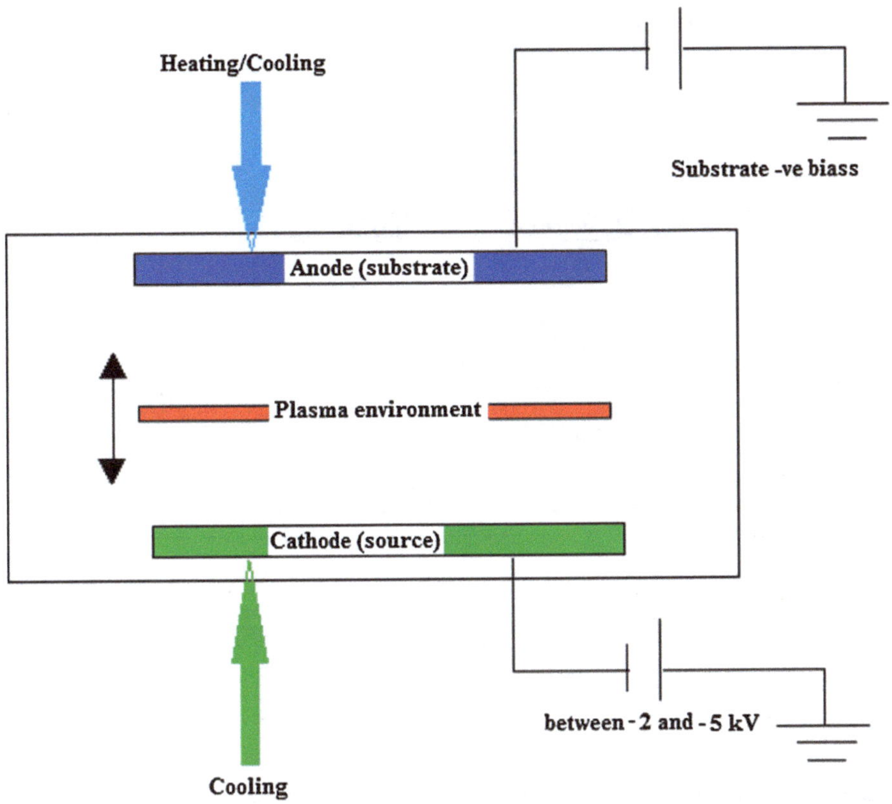

Fig. 3.3 Schematic diagram of the DC sputtering process

to smaller substrate sizes [8, 9]. The major advantages of RF sputtering include neutralization of the positive charge built up by the alternating potential. It is easier to maintain the plasma, since it can operate at lower inert gas pressures (1–15 m Torr). There are fewer gas collisions; hence, there is more line-of-sight deposition. This works well with insulating target materials, and it has a higher level of efficiency.

Magnetron sputter deposition uses powerful magnets to confine the discharged plasma into the region that is very close to the target or source material. This will increase the possibility of ionizing the collisions with the argon gas molecules. These would, in turn, help to improve the deposition rate—by maintaining a higher density of ions in the system.

The magnetron sputtering is used with either a DC or a RF power source. It uses the combination of electricity and magnetic fields. It helps to increase the ionization of the inert gas. Higher sputter rates can be achieved at lower inert gas pressure levels, and hence, fewer gas collisions that would result in more line-of-sight by increasing the probability of electrons striking the inert gas atoms [12]. It also

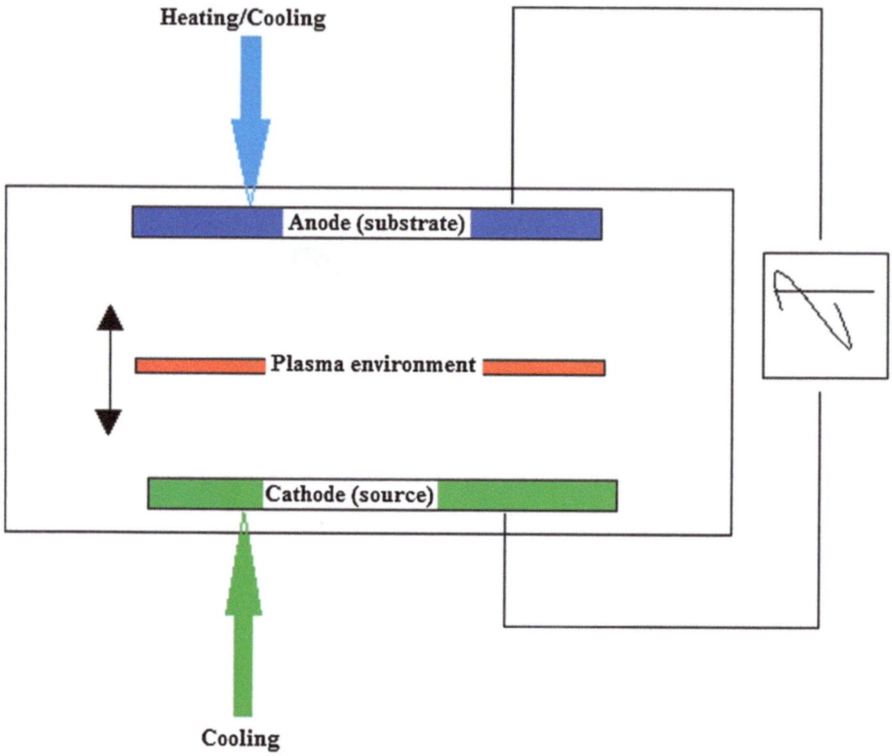

Fig. 3.4 Schematic diagram of RF sputtering system

increases the electron path length; it traps the electrons near the source material; and it causes more ionization near the cathodes, which results in fewer electrons reaching the substrate; and hence, it produces less heat.

Some of the advantages of the magnetron sputtering deposition process includes the very high deposition rate of about ten times higher than the conventional sputtering system; it reduces the electron bombardment on the substrate, thereby producing less substrate heating (the deposition temperature is about 100°–150 °C, thereby extending the operating vacuum range) and the ability to operate at lower pressures of less than 0.1 Pa. Some of the disadvantages of this process include the following: It causes an erosion track in the target material that leads to poor efficiency of the sputtering yield versus the target volume when compared with the non-magnetron sputtering system; it produces non-uniform removal of particles from the target material, and this causes the deposition thickness of the substrate to be uneven.

Reactive sputtering is another sputtering process that is used to produce compound thin-film coatings through the introduction of reactive gases that are introduced into the system by mixing the reactive gases with the inert gas. The compound is formed through the reaction of the sputtered atoms and the reactive gases in the plasma region, which upon reaching the substrate, the compound is deposited. The reaction can also take place at the source material surface, as well as at the surface of the substrate. The main disadvantage of this process is that if the supply of the reactive gas is too high, the poisoning of the source material surface would occur; and this may adversely affect the deposition rate. There is a need for proper adjustment of the reactive gas flow rate, in order to achieve a good stoichiometry.

The quality of the deposited film depends largely on the deposition conditions. These include the substrate temperature, the deposition pressure, and the power, as well as the inert gas and the reactive gas flow rate. The coatings produced by the reactive sputtering process are used in applications, such as the food-packaging barrier coatings, to prevent water and oxygen vapour transmission in food preservation, multilayer oxide anti-reflective coatings, scratch-resistant coatings, multilayer oxide filters for lasers, oxide and nitride coatings used for decorative purposes, wear-resistant coatings on drill bits, mills, and automotive components, such as piston rings, shafts, and gears.

Additionally, nitride coatings for tools and oxy-nitride coatings for solar absorbent layers are used in a thermal solar cell.

3.2.3 The Advantages of Sputtering Deposition Compared to Vacuum Evaporation

Sputtering deposition techniques are more versatile than vacuum evaporation—with the possibility of depositing a wide variety of insulators, metals, alloys, and compounds. The choices of materials are limited in the vacuum evaporation system. The sputtering target or source material provides a stable vaporisation source. Furthermore, the reactive deposition can be easily achieved using reactive gases, which are activated in plasma. A small-volume deposition chamber can be utilized, which allows the target material and the substrate to be spaced as close together as possible. The replication of the target composition in the deposited film can readily be achieved, but, even if the materials have widely different sputter yields, this cannot be easily achieved in the vacuum evaporation system. The capability of in situ cleaning of the substrate before the deposition process can be achieved through the 'back-sputtering' process.

This is done by reversing the system polarity in the DC sputtering system that is used to clean the substrate surface in situ, which has a great influence on the electrical and the mechanical properties of the deposited material on the substrate, as well as the deposited substrate's adhesion quality. Self-sustained glow discharge is achievable, and this can be created by the breakdown of the inert gas. The neutral atoms that are ejected from the cathode diffuse to the anode or the chamber walls, thereby allowing a better coverage and better adhesion from the ion mixing, which arises from the energy spread.

The sputtering deposition process is a low-vacuum process, and it is a lower line-of-sight deposition process. Smaller grain sizes with many orientations are achievable in the sputtering deposition system. The possibility of material decomposition is very low, and scale up is easy to achieve. The shadowing effect is also very low, and the control of the properties can easily be achieved.

3.2.4 *Disadvantages of Sputter Deposition Over Vacuum Evaporation*

A low sputtering rate results in a low deposition rate compared with the thermal evaporation system. Most of the incident energies are converted to heat, and this needs to be removed from the system. In the reactive sputter deposition process, the gas composition must be well-controlled, in order to prevent the poisoning of the sputtering target. Sputtering can damage the surface of the substrate—because of the high-impact energy of the sputtering gas that results in high substrate heating. This renders it unsuitable for the deposition of material on gate oxide.

Sputtering deposition is a high-energy process, with more than 10 eV. Kinetic energy when compared with the vacuum evaporation process emits energy as low as 0.5 eV. Gas inclusions in the deposit are more common with the sputtering deposition system, because of the low vacuum that is involved. It is also very expensive to replace the source material with the sputtering deposition system, while that can easily be achieved with the evaporation system. The capital investment is much higher in the sputtering system compared to the evaporation system.

3.2.5 *Plasma Spray–Physical Vapour Deposition System*

The plasma spray–physical vapour deposition (PS-PVD) process was developed recently by Sulzer-Metco. The PS-PVD process is a hybrid deposition process. It combines a cost-effective low-pressure traditional plasma-spraying (LPPS) technique, which enables the evaporation of the coating materials and the PVD technique. A typical standard LPPS process is conducted at a pressure between 50 and 200 mbar, and it permits the deposition of 20-μm–1-mm-thick coatings [16]. By lowering the operating pressure, the plasma flame can be as long as 50–500 mm, which allows the deposition of a uniform and homogeneous coating.

In the PS-PVD process, the operating pressure is in the range of 0.5–2 mbar, and the plasma flame is as long as over 2 metres, with the flame diameter reaching between 200 and 400 mm. The pressure in the PS-PVD process is higher than in the PVD process, but the higher plasma stream velocity, which is above 2000 m/s, and the high plasma stream temperature enable the feedstock to be easily vapourised.

This, in turn, permits the deposition of coatings—even in places, which are not inside the stream [16].

The areas of application of the physical vapour deposition process are presented in the next subsection.

3.2.6 Areas of Application of Physical Vapour Deposition Process

PVD is used in the manufacture of coatings on products that require thin-film coatings and functionally graded coatings for improved mechanical, optical, chemical, or electronic functionality. The areas of application include the coatings on semiconductor devices, aluminized polyethylene terephthalate (PET) film for food packaging, and coated cutting tools for metalworking. Industries, such as the automotive, aerospace, biomedical, defence, die, and moulding industries, are using the PVD process for the production of various forms of coatings. The PVD process is also used for wear- and abrasive-resistant coating on forming and cutting tools. The chemical vapour deposition process is discussed in the next section.

3.3 Chemical Vapour Deposition

The chemical vapour deposition (CVD) process is another type of deposition process that can be used to produce thin-film deposition and functionally graded coating. The chemical vapour deposition process is achieved by placing the material to be coated inside a vacuum; the coating material is then vapourised—either by heating the material, or by reducing the pressure around the material until the material vaporises. The precursor gas or gases are then introduced into the chamber containing the heated objects to be coated [16, 17].

Chemical reactions occur on and near the heated surface of the substrate. This results in the deposition of a thin film on the surface. The reactions of by-products are exhausted out of the chamber with some unused precursor gases. The schematic diagram of the CVD process is shown in Fig. 3.5. The process is carried out in hot-wall reactors and cold-wall reactors, at pressures above the atmospheric pressures and at temperatures ranging between 200° and 1600 °C.

There are different types of enhanced CVD processes available incorporating processes, such as the plasmas, the ions, the photons, the lasers, the hot filaments, or the combustion reactions that can be incorporated into the CVD process to help increase the deposition rates and/or to lower the deposition temperatures [17]. There are also many derivatives of the CVD processes, which include the metal organic chemical vapour deposition (MOCVD) process, the organometallic chemical vapour deposition (OMCVD) process, the organometallic vapour-phase epitaxy

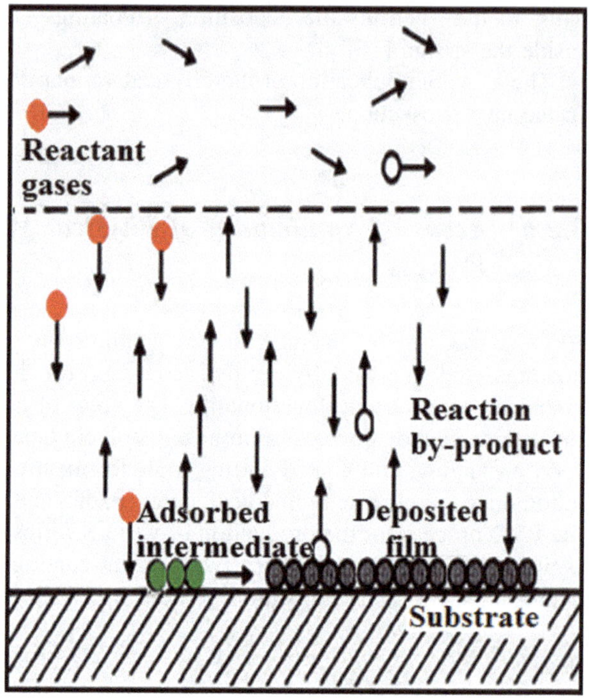

Fig. 3.5 Schematic diagram of chemical vapour deposition process

(OMVPE) process, and the metalorganic vapour-phase epitaxy (MOVPE) process [17]. The advantages of the CVD process include the following: the insides and the undersides of geometrical features are easily covered in the coating process; and high-aspect ratio holes and other features can be completely filled, in contrast to the PVD process, which comprises a line-of-sight between the surface to be coated and the source material. A wide variety of materials can be deposited, using the CVD process—and with very high levels of purity. This is achieved as a result of the ease with which the impurities are removed from the process. Relatively high deposition rates can be achieved, and the system does not require as high a vacuum as the PVD processes do.

3.3.1 Advantages and Disadvantages of the Chemical Vapour Deposition Process

The chemical vapour deposition process does not require vacuum or high levels of electrical power. The substrate does need to be rotated to achieve wide-step coating. The operating cost is very low, because of the low vacuum that it employs. It is suitable for coating certain intricacies, such as holes and slots. Some of the disadvantages of the CVD process include the following: the precursors need to be volatile at near-room temperatures. CVD precursors can also be highly toxic, explosive, and corrosive. The by-products of the CVD reactions can be highly hazardous. Some of the precursors are very expensive. The films are usually deposited at elevated temperatures. This puts some restrictions on the kind of substrates that can be coated using the CVD process. The CVD process produces thermal stresses in the deposited films when the substrate material and the deposited material have different thermal coefficients of expansion.

The coating with several metals using the CVD process is not possible.

3.3.2 Applications of the Chemical Vapour Deposition Process

The early application areas of the CVD process include the wear- and corrosion-resistant coatings, and in the fabrication of the structural components. Most of the recent applications of the CVD process are for thin-film deposition and functionally graded coatings for tribological applications, for the protective coatings for tool steels. The use of the CVD process for the production of diamond and diamond-like carbon (DLC) coatings has significant commercial applications, because of its high degree of hardness, thermal conductivity, chemical inertness, and electronic properties [17]. The CVD process is used in the fabrication of solid-state electronic devices, ball bearings, cutting tools, and nuclear reactor components. There are other types of fabrication processes that can be used to produce functionally graded coatings—apart from those that have already been discussed in this chapter. Processing techniques for the production of thin-film functionally graded materials are summarized in Fig. 3.6.

The fabrication process for the bulk FGM is presented in the next section.

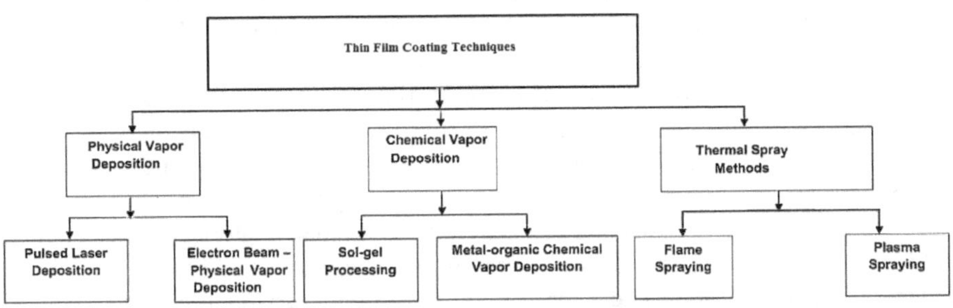

Fig. 3.6 The thin-film coating processes

3.4 The Fabrication Process of Bulk Functionally Graded Materials

The fabrication processes of thin-coating functionally graded materials were discussed in the previous section; however, some of the applications where FGMs are used require bulk functionally graded materials, because of the extreme nature of the working environments. Some of fabrication technologies for producing bulk functionally graded materials, such as the powder metallurgy, the centrifugal casting method, and the tape casting method, are explained in this section, together with their merits, demerits, and areas of application.

3.4.1 Powder Metallurgy

The powder metallurgy (PM) process is an old manufacturing process for making engineering parts, which is also now used to produce functionally graded materials. The series of operations that are involved in the production of FGMs include the preparation of the powder materials, the processing of the powder, the forming operations, and the sintering or pressure-assisted hot consolidation—depending on the service requirement of the functionally graded materials being produced. Powders of metals, alloys, compounds, and ceramic materials are readily available from the powder manufacturers in a range of particle sizes and distributions. The graded materials are built up using the powder particles of the constituent materials. The powder particles are used as building blocks using the powder mixtures with changing average particle sizes or composition, depending on the path dictated by the design of the functionally graded materials.

By stacking the graded powder, a stepwise variation in composition is achieved in the green part that is produced. It is called a green part because the structure produced is fragile and it needs to undergo a consolidation process. After the production of the graded green part, the consolidation of the green parts is achieved through sintering or hot pressing. Sintering is the process of heating the green part in a furnace to cause some of the constituent materials of the FGM to be melted or surface melted. This helps to bond the particles together. The sintering process requires high temperatures that can promote proper densification, as well as diffusion processes and chemical reactions to take place. Also, the thermodynamic factors during the sintering process can also be used to design the functionally graded materials. For example, during the liquid-phase sintering process, it is possible to produce porosity-graded functionally graded materials, which can be achieved using powder mixtures of different particle sizes and shapes. Further, the chemical composition gradient or the microstructure gradient can be achieved during the sintering process. All these steps involved in the powder metallurgy process are explained in the following subsections.

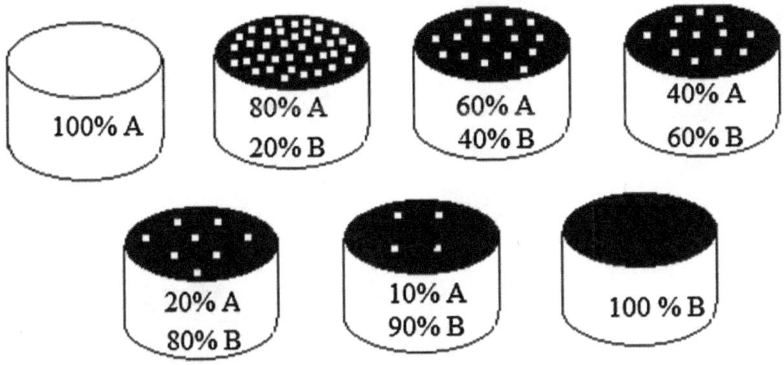

Fig. 3.7 Powder mixture preparation

3.4.1.1 Formation of the Graded Powder

The powder mixtures are prepared, based on the functionally graded material design, where the powder mixtures are set aside. For example, for a gradient that requires two powders A and B, 10% of powder A is mixed with 90% of powder B, using the appropriate mixing process for the recommended length of time and set aside and labelled appropriately. The second powder mix is prepared using 20% of powder A and 80% of powder B. It is also labelled appropriately, and it is then set aside. All the powder mix ratios are prepared in this way, labelled, and set aside. The schematic diagram of this process is presented in Fig. 3.7. After the various powder mixtures have been prepared and set aside, as shown in Fig. 3.7, the next step is to stack the powders in a die, and this process is explained in the next subsection.

3.4.1.2 Powder Stacking in the Powder Metallurgy Process

In the previous example, the first powder to be stacked could be 100% of powder A, followed by 90% of powder A plus 10% of powder B, then 80% of powder A plus 20% of powder B, 70% of powder A plus 30% of powder B, 60% of powder A plus 40% of powder B, and so on, until finally, the 100% of powder B has been added to the stack. As each of the powder mixtures is placed in the die, the powder mixture is compacted using a ram with appropriate pressure applied. The schematic diagram of the stacked powder mixture is shown in Fig. 3.8.

After the stacking process and the ramming operations, the next step is sintering. The process of sintering is presented in the next subsection.

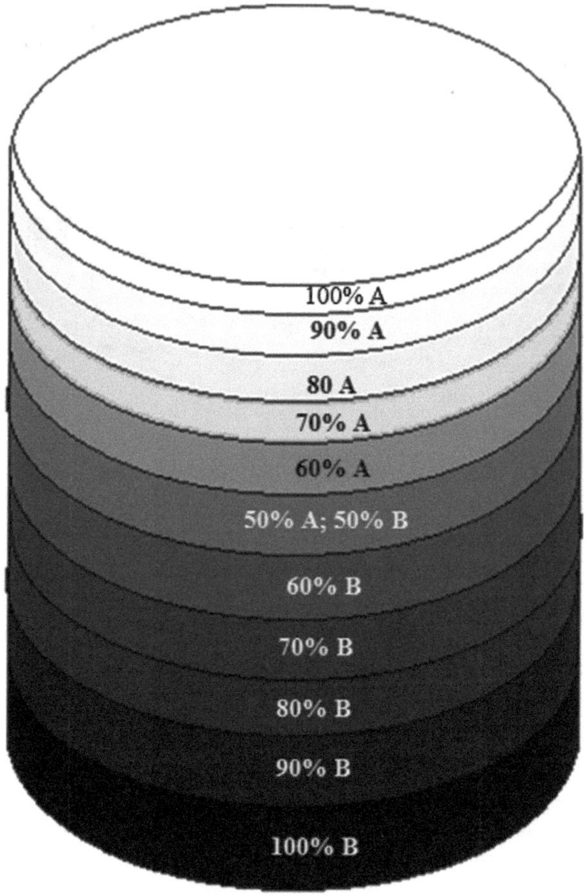

Fig. 3.8 Stacked powder mixture

3.4.1.3 Sintering

After the powder stacking and the ramming process are completed, the green part that is produced is placed in a furnace at a temperature below the melting temperature of the constituent material, with the lowest melting temperature. The temperature is sufficiently high to cause the surface melting of the powder with the lowest melting temperature. The melting of the surface of this powder will cause the proper bonding of the powder together, and with high strength. The sintering process can be performed simultaneously with the application of pressure, which is known as the hot-pressing process. This is desirable when a higher density is required in the part being made. In the cold-pressing process, the sintering process is performed without the application of pressure, but the powders were compacted using sufficiently high pressure during the stacking process. There are three different types of sintering processes, namely the electric furnace sintering, the high frequency induction heating, and the spark plasma sintering (SPS). The SPS is a newly developed sintering process that uses plasma for the sintering process [21]. The advantages of using the SPS systems include the following: the process is faster, and the ease of operation is enhanced with low running cost. The flow chart of the powder metallurgy process is shown in Fig. 3.9. The advantages and the disadvantages of the powder metallurgy process for the production of FGM are presented in the next section.

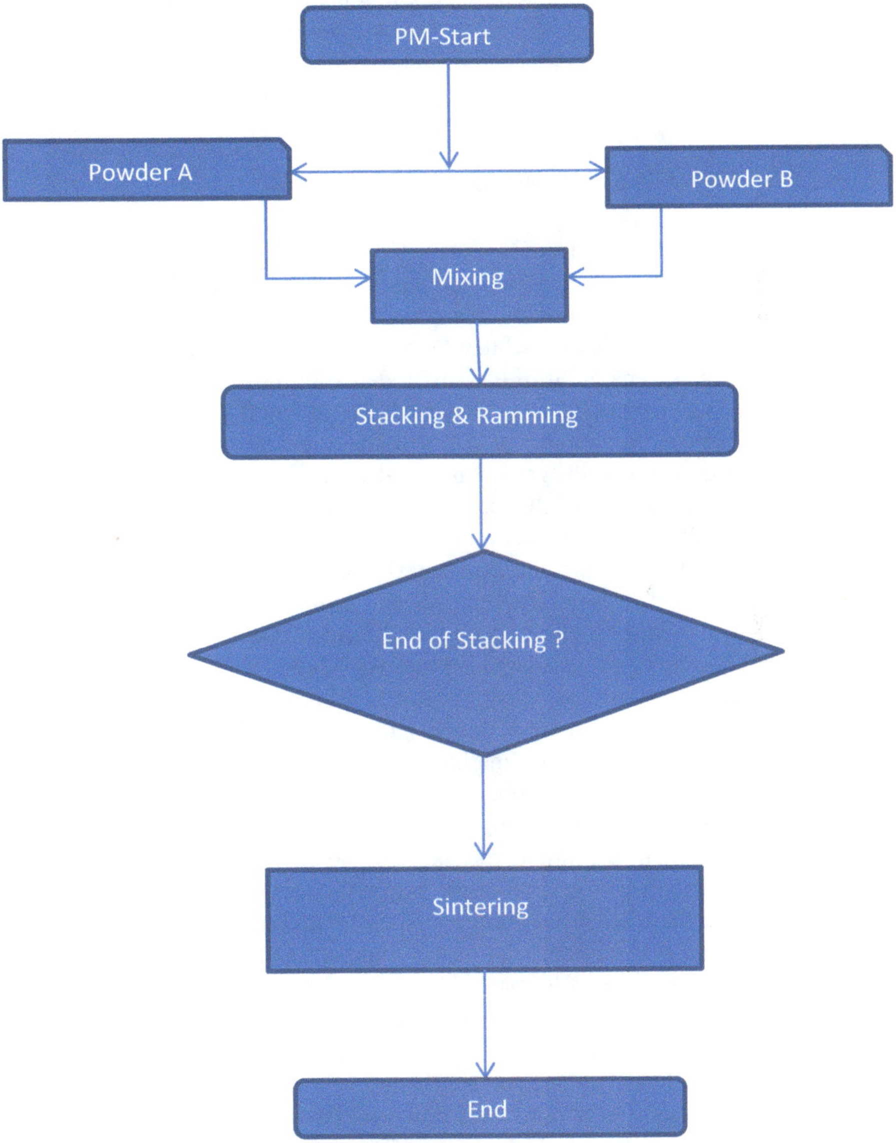

Fig. 3.9 Flow chart of powder metallurgy for production of FGM

3.4.1.4 The Advantages and Disadvantages of Powder Metallurgy

Powder metallurgy is cheap, and it is easy to operate. A wide control of composition and microstructure is achievable with near net-shape forming capability. A wide range of materials can be processed through this route. The energy consumption in the powder metallurgy process is low, and the processing time is short. It permits mass production with controllable properties. The main disadvantages of powder metallurgy process are that the strength of the part is limited and that highly intricate parts cannot be produced by this process. The centrifugal casting method is explained in the next section.

3.4.2 *The Centrifugal Casting Method*

The centrifugal casting method is achieved by pouring a molten material that contains another reinforcing material—either in molten, or in solid state—into a mould inside a rotating die to produce a functionally graded material. By rotating the die, a centrifugal force is created that helps to draw the molten material towards the mould and create separation in the suspended solid powder material—and the melting of the two materials, as a result of the different densities of the two materials, and hence the creation of a functionally graded material. The graded distribution of the functionally graded material formed by the centrifugal casting method would be significantly influenced by the processing parameters, such as the difference in density between the reinforcing powder particles and the molten material, the particle size and the particle size distribution of the powder, the viscosity of the molten material, and the solidification time.

The centrifugal casting method is one of the most effective methods for processing the bulk of the functionally graded materials due to its wide range control of composition and microstructure. There are two types of fabrication of a FGM when using the centrifugal casting method, namely the centrifugal solid-particle method (CSPM) and the centrifugal in situ method (CISM).

In the CSPM method, the melting point of the reinforcement material is significantly higher than the processing temperature, and the reinforcing powder particles remain solid in the liquid matrix. This helps to produce selective reinforcement at the surface of the component, while producing a core that is more of the matrix material. The selective reinforcement of the component surface helps to produce a high wear-resistant material in the outer surface, while maintaining high toughness at the core of the bulk material.

A steeper compositional gradient is produced with the CSPM method because the motion of the solid particles during the rotation under the centrifugal force is governed by Stoke's law. This means that the larger the particle size, the larger the migration distance [22]. The processing parameters controlling such process are the particle size and the distribution of the reinforcing material, the volume fraction of the matrix and that of the reinforcing material, the viscosity of the matrix, the cooling rate, the solidification time, and the speed of rotation of the die [23].

All these process parameters have a significant influence on the properties of the FGM composite produced.

In the CISM method, the melting point of the reinforcing particles is lower than the processing temperature, and the centrifugal force is applied during the solidification process, which is similar to the production of in situ composites using the crystallization phenomenon. The density difference between the matrix and the reinforcing particles causes the partial separation of the materials in the liquid state. The gradient composition is formed before the crystallization of the primary crystals. The primary crystals in the matrix are formed on the basis of the local chemical composition, and it is precipitated because of the density difference, which further produces an additional compositional gradient [24].

The density of the surface part of the functionally graded material is higher than that of the inner part with this CISM method. The schematic diagram of the centrifugal casting method is shown in Fig. 3.10. The advantages and disadvantages of centrifugal casting will be presented in the next subsection.

3.4.2.1 Advantages and Disadvantages of the Centrifugal Casting Method

The main advantage of using the centrifugal casting method for the production of functionally graded materials is that a continuous gradient can be produced using the centrifugal casting process. The main disadvantages of the centrifugal casting method include the following: It can only be used to produce a cylindrical section, such as tube, bushing, and cylindrical or tubular castings that are simple in shape. The gradation is limited by the centrifugal force and the density difference of the constituent materials [5, 25, 26]. The tape casting method is discussed in the next subsection.

3.4.3 The Tape Casting Method

The tape casting process is achieved by the spreading of a slurry mixture onto a moving belt, and then passing the moving belt under the blade edge, in order to shape the slurry into a tape of constant thickness. The schematic diagram of the tape casting process is shown in Fig. 3.11.

The slurry mixture is formed by putting the required powder mixture into an organic solvent with suitable binders and plasticisers. The slurry is then cast into a film thickness to form a tape shape with the size range between several μm to few mm thick by the casting blade [27]. The solvent is then dried off, leaving a green part behind. Stepped gradients of FGM are produced by stacking the tapes of different compositions. The stack of tapes that is produced is then sintered under an

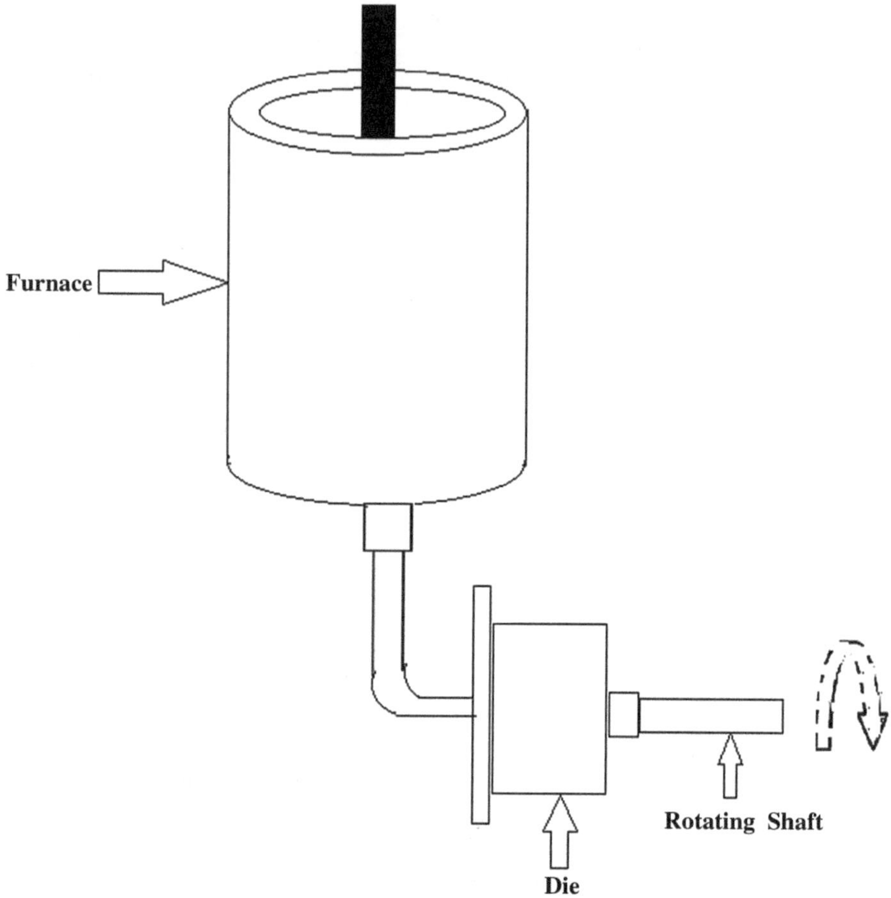

Fig. 3.10 Schematic diagram of the centrifugal casting method

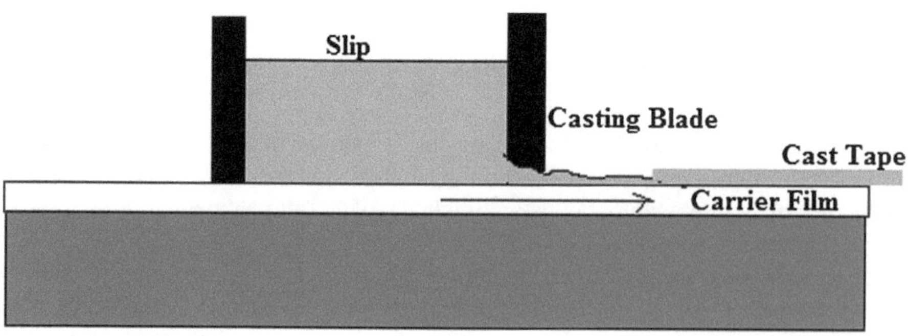

Fig. 3.11 Schematic diagram of the tape casting process

elevated temperature from 50° to 200 °C and a pressure of about 3–30 MPa. The sintering of the stacked graded part at elevated temperature and pressure would help to remove the organic binder, and also help to increase the density of the part. The main advantage of the tape casting method is that it helps in the production of the high-resolution functionally graded materials. The main disadvantage of this process is that the strength of the part produced is limited and it depends on the sintering temperature and pressure. The slip casting method is also another casting method that has been used to produce functionally graded materials. The readers can consult the following references for further reading [28, 29]. The additive manufacturing technology, which is an advanced manufacturing process, is also used for the production of functionally graded materials. This will be discussed in detail in the next chapter.

3.5 Summary

The fabrication methods for the thin functionally graded coating and the bulk functionally graded materials have been presented in this chapter. The physical vapour deposition process and the chemical vapour deposition process are the two main fabrication methods for the production of functionally graded coating that are well established and used at commercial levels. These manufacturing processes have been reviewed in this chapter with their merits, demerits, and their areas of application highlighted. The powder metallurgy and the casting-based fabrication technologies for the production of bulk functionally graded materials have also been discussed in detail. The next chapter focuses on the fabrication of the functionally graded materials using the additive manufacturing technologies.

Acknowledgements This work is supported by the University of Johannesburg Research Committee Fund, the Department of Higher Education and Training (DHET) South Africa, the National Laser Centre Rental Pool Programme (RPP) contract number NLC-LREHA02-CON-001 and L'Oreal-UNESCO For Women in Science.

References

1. Mahamood, R.M., Akinlabi, E.T., Shukla, M., Pityana S.: Functionally graded material: an overview. In: Proceedings of the World Congress on Engineering WCE 2012, vol. 3, pp. 1593–1597 (2012)
2. Dumont, A.-L., Bonnet, J.-P., Ferreira, J.M.F., Chartier, T.: MoSi2/Al2O3 FGM: elaboration by tape casting and SHS. J. Eur. Ceram. Soc. **21**, 2353–2360 (2001)
3. Saiyathibrahim, A., Mohamed, N.S.S., Dhanapal, P.: Processing techniques of functionally graded materials—a review. In: International Conference on Systems, Science, Control, Communication, Engineering and Technology, pp. 98–105 (2015)

4. Jamaludin, S.N.S., Mustapha, F., Nuruzzaman, D.M., Basri, S.N.: A review on the fabrication techniques of functionally graded ceramic-metallic materials in advanced composites. Acad. J. **8**(21), 828–840 (2013)
5. Kieback, B., Neubrand, A., Riedel, H.: (2003). Processing techniques for functionally graded materials. Mater. Sci. Eng. **362**, 81–105
6. Håkansson, G., Hultman, L., Sundgren, J.-E., Greene, J.E., Münz, W.-D.: Microstructures of TiN films grown by various physical vapour-deposition techniques. Surf. Coat. Technol. **48** (1), 51–67 (1991)
7. Reichelt, K., Jiang, X.: The preparation of thin films by physical vapour-deposition methods. Thin Solid Films **191**(1), 91–126 (1990)
8. Liu, C., Leyland, A., Bi, Q.: A Matthews, Corrosion resistance of multi-layered plasma-assisted physical vapour-deposition TiN and CrN coatings. Surf. Coat. Technol. **141**(2–3), 164–173 (2001)
9. Wiiala, U.K., Penttinen, I.M., Korhonen, A.S., Aromaa, J., Ristolainen, E.: Improved corrosion resistance of physical vapour-deposition coated TiN and ZrN. Surf. Coat. Technol. **41**(2), 191–204 (1990)
10. Aubert, A., Gillet, R., Gaucher, A., Terrat, J.P.: Hard chrome coatings deposited by physical vapour deposition. Thin Solid Films **108**(2), 165–172 (1983)
11. Ong, J.L., Lucas, L.C., Raikar, G.N., Weimer, J.J., Gregory, J.C.: Surface characterization of ion-beam sputter-deposited Ca-P coatings after in vitro immersion. Colloid Surf. **87**, 151–162 (1994)
12. Wolke, J.G.C., van Dijk, K., Schaeken, H.G., de Groot, K., Jansen, J.A.: Study of the surface characteristics of magnetron-sputter calcium phosphate coatings. Biomed. Mater. Res. **28**, 1477–1484 (1994)
13. van Dijk, K., Schaeken, H.G., Wolke, J.G.C., Maree, C.H.M., Habraken, F.H.P.M., Verhoven, J., Jansen, J.A.: Influence of discharge power level on the properties of hydroxyapatite films deposited on Ti6Al4V with RF magnetron sputtering. J. Biomed. Mater. Res. **29**, 269–276 (1995)
14. van Dijk, K., Schaeken, H.G., Wolke, J.G.C., Jansen, J.A.: Influence of annealing temperature on RF magnetron-sputtered calcium phosphate coatings. Biomaterials **17**, 405–410 (1996)
15. Wolke, J.G., van der Waerden, J.P., Schaeken, H.G., Jansen, J.A.: In vivo dissolution behaviour of various RF magnetron-sputtered Ca-P coatings on roughened titanium implants. Biomaterials **24**, 623–629 (2003)
16. Zhou, Z.-J., Song, S.-X., Du, J., Zhong, Z.-H., Ge, C.-C.: Performance of W/Cu FGM-based plasma facing components under high heat load test. J. Nucl. Mater. **363**, 1309–1314 (2007)
17. Kumar, V., Kandasubramanian, B.: Processing and design methodologies for advanced and novel thermal barrier coatings for engineering applications. Particuology **27**, 1–28 (2016)
18. El-Wazery, M.S., El-Desouky, A.R., Hamed, O.A., Mansour, N.A., Hassan, A.A.: Preparation and mechanical properties of zirconia/nickel functionally graded materials. Arab. J. Nucl. Sci. Appl. **45**(2), 435–446 (2012)
19. Nemat-Alla, M.M., Ata, M.H., Bayoumi, M.R., Khair-Eldeen, W.: Powder-metallurgical fabrication and microstructural investigations of aluminum/steel functionally graded material. Mater. Sci. Appl. **2**, 1708–1718 (2011)
20. Jin, X., Wu, l., Sun, Y. and Guo, L.: Microstructure and mechanical properties of ZrO_2/NiCr functionally graded Materials. Mater. Sci. Eng. A **509**, 63–68 (2009)
21. Menga, F., Liua, C., Zhangb, F., Tiana, Z., Huanga, W.: Densification and mechanical properties of fine-grained Al_2O_3–ZrO_2 composites consolidated by spark-plasma sintering. J. Alloy. Compd. **512**, 63–67 (2012)
22. Dusenbery, D.B.: Living at Micro Scale. Harvard University Press, Cambridge, Mass (2009)
23. Watanabe, Y., Sato, H., Ogawa, T., Kim, I.-S.: Density and hardness gradients of functionally graded material ring-fabricated from Al-3 mass%Cu alloy by a centrifugal in-situ method. Mater. Trans. **48**(11), 2945–2952 (2007)

24. Watanabe, Y., Sato, R., Kim, I.-S., Miura, S., Miura, H.: Functionally graded material fabricated by a centrifugal method from ZK60A magnesium alloy. Mater. Trans. **46**(5), 944–949 (2005)
25. Zygmuntowicz, J., Miazga, A., Konopka, K., Edrysiak, K.J., Kaszuwara, W.: Alumina matrix ceramic-nickel composites formed by centrifugal slip casting. Process. Appl. Ceram. **9**(4), 199–202 (2015)
26. EL-Wazery, M.S., EL-Desouky, A.R.: A review on functionally graded ceramic-metal materials. Mater. Environ. Sci. **6**(5), 1369–1376 (2015)
27. Liu, S., Shen, Q., Luo, G., Li, M., Zhang, L.: Fabrication of W/Cu FGM by Aqueous-tape casting. J. Phys. Conf. Ser. **419**, 1–5 (2013)
28. He, X., Du, H., Wang, W., Jing, W., Liu, C.: Fabrication of ZrO2-SUS functionally graded materials by slip casting. Key Eng. Mater. **368–372**, 1823–1824 (2008)
29. Neirinck, B., Mattheys, T., Braem, A., Fransaer, J., Biestand, O.V., Vleugels, J.: Preparation of titanium foams by slip casting of particle-stabilized emulsions. Adv. Eng. Mater. **11**(8), 633–636 (2009)

Chapter 4
Additive Manufacturing of Funtionally Graded Materials

4.1 Introduction

According to the ASTM F42 technical committee on additive manufacturing (AM) standards, additive manufacturing is defined as the process of making three-dimensional objects (3D) from the 3D model data of the object through the addition of materials layer-upon-layer. This is different from the subtractive manufacturing systems that involve material removal [1]. The AM process is also known as 3D-printing, direct-digital manufacturing, layered manufacturing, solid-free form fabrication, rapid prototyping, and rapid manufacturing. Additive manufacturing is used for the production of prototypes for form and fit checking and design verifications, rapid tooling and pattern production, as well as the rapid manufacturing of functional parts. Additive manufacturing technologies allow the fabrication of 3D geometrically complex parts—without the need of special jigs and fixtures that are used in the subtractive manufacturing technology.

The AM process is very attractive; because of the numerous advantages it possesses, such as significantly shortening the product lead time; a complex part can be built as a single unit object, as against being broken down into a number of smaller parts, when the conventional manufacturing process is used. These are cost-effective, customized parts that can be produced at a mass production rate with lower cost [2]. This process also includes the possibility of making parts that comprise functionally graded materials [3, 4].

There is a wide variety of AM technologies that are recently being grouped into seven classes by the ASTM F42 technical committee [1]. The seven classes are: Vat Photopolymerization, Material Jetting, Binder Jetting, Material Extrusion, Powder-Bed Fusion, Sheet Lamination, and Directed-Energy Deposition. Examples of each of the seven classes of the AM technologies, respectively, are: Stereolithography (SLA) [5]; Multi-jet modelling (MJM) [6]; plaster-based 3D printing (PP) [7]; Fused-Deposition.

R.M. Mahamood and E. Titilayo Akinlabi, *Functionally Graded Materials*,
Topics in Mining, Metallurgy and Materials Engineering,
DOI 10.1007/978-3-319-53756-6_4

Modelling (FDM) [8]; Selective-Laser Sintering (SLS) [9]; Laminated Objective Manufacturing (LOM) [10]; and the laser-metal deposition (LMD) process [11]. Some of these classes are used for prototyping and the production of casting patterns; they are: Vat Photopolymerization; Material Jetting; and Binder Jetting. Only four classes, out of the seven classes of the AM technologies, are used for the production of functional parts made of metals, alloys, and composite materials. These four classes that are used for the production of functional parts are also used for the fabrication of parts that are made of FGM.

They comprise: Material Extrusion; Powder-Bed Fusion; Directed-Energy Deposition; and sheet lamination. These four classes of the AM technologies are explained in detail in this chapter. Each of these classes is described, according to their use for the production of FGMs. Some of the research works in this field are also reviewed and presented in this chapter.

The use of the MJM and plaster-based 3D printing for the production of prototypes and functional parts of polymer–polymer, or polymer–ceramic, ceramic–ceramic FGMs are reported in the literature; but they are not discussed in detail in this chapter.

4.2 Material Extrusion

Fused-deposition modelling (FDM) is a trademark of Stratasys company, and it belongs to the material-extrusion class of AM processes. There are five basic steps that are involved in any AM technology; and these are as follows: The receiving of the CAD file into the AM machine; the conversion of the CAD file into the standard triangulation language (STL), or the additive manufacturing file (AMF); the third stage is the slicing of the converted file into two-dimensional (2D) layers of a triangle, representing the 3D CAD data; the fourth step is the layer-by-layer building up of the part following the path dictated by the 2D sliced CAD data; finally, the part removal from the machine and the performance of finishing operations, such as support-structure removal and heat treatment.

The flow chart of this process is shown in Fig. 4.1. It is the building process that differentiates between one class of the AM process and the other. For the material-extrusion class of AM, the building process is achieved, according to the following steps:

- The nozzle deposits the extruded material onto the building platform.
- Based on the cross-sectional area of the sliced 2D data, the perimeter of the sliced data is first deposited; and the shape is filled up, according to the raster pattern chosen.
- After the first layer is completed, the building platform is lowered by one layer of thickness and the next layer is added, as done in the previous step. The previous layers are fused together with the successive layers because the deposited materials are still in a melted state.
- The process is repeated until the building process has been completed.

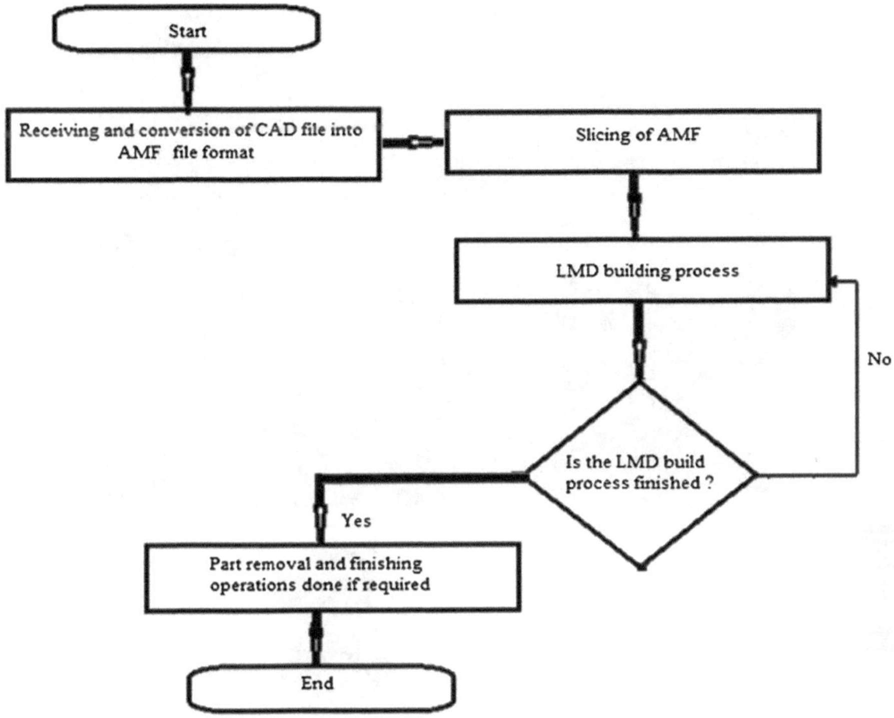

Fig. 4.1 Flow chart of the stages of the laser-metal deposition process

The FDM uses a wire-like filament material to build up the three-dimensional object, layer-by-layer. The material (filament in a spool) is heated and drawn through the extruder nozzle; then, it is deposited layer-after-layer, as shown in Fig. 4.2a. The schematic diagram of the FDM process is shown in Fig. 4.2b.

The process uses thermoplastic materials, such as ABS, polycarbonate, PC/ABS, polyphenylsulfone, and ULTEM 9085. These materials harden quickly upon reaching the building platform. The platform is lowered after each layer is deposited. The layer thickness is equivalent to the distance the building platform is lowered. The overhangs and support structures used are usually made up of water-soluble materials, which would make their removal easy—after the building process.

This AM technology is most commonly found in the domestic 3D printers [12]. FDM is used to produce prototypes, functional parts, and parts made with FGMs. The most important process parameters in this process are: the filament width; the raster-fill angle and the raster-fill pattern. All these process parameters have an influence on the properties of the produced objects [13–16]. The extrusion pressure must also be kept at a constant velocity, in order to produce a good deposit with a

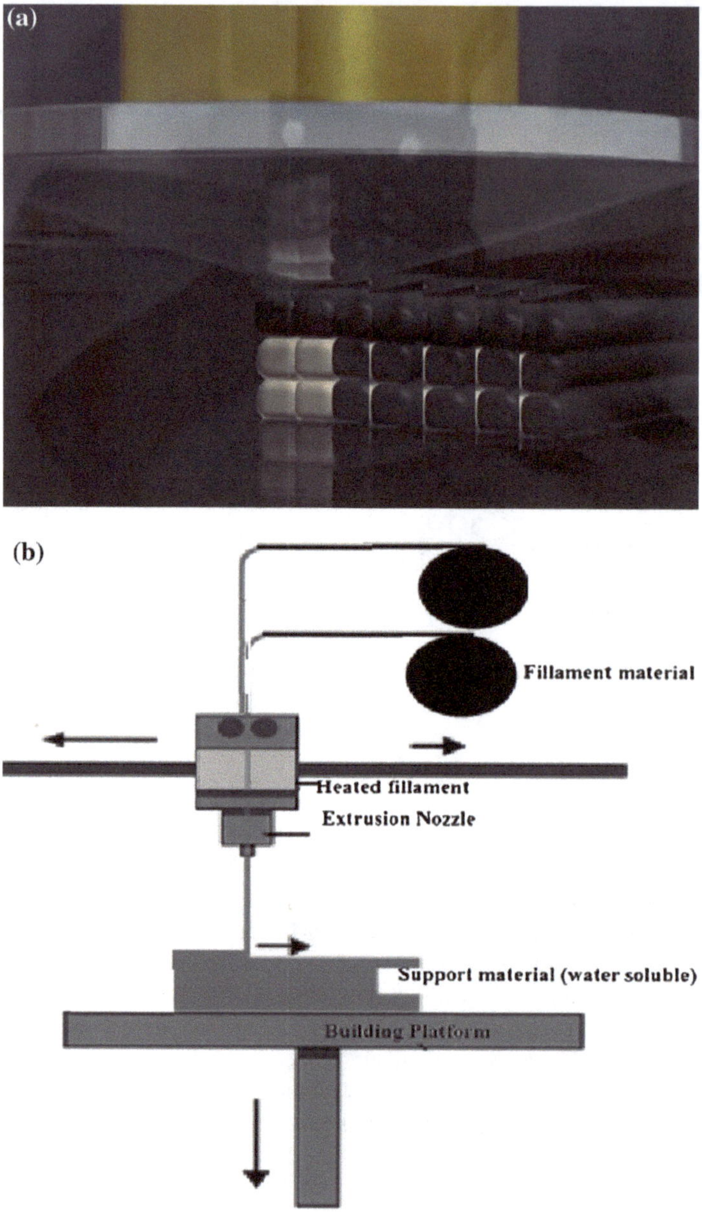

Fig. 4.2 Schematic diagram showing: **a** material being extruded through the nozzle; and **b** the FDM process

proper surface finish. The main advantages of the material-extrusion process are that the materials are cheap and readily available. The limitation of the process includes: The part quality is limited by the nozzle radius; the accuracy of the final part is limited by the material's nozzle thickness; the pressure of the extruded material affects the quality of the surface finish; and the process is slow in comparison to other processes.

Freeze-form Extrusion Fabrication (FEF) is another material-extrusion class process [17]. The process is used to build a 3D object layer-after-layer through the computer-based controlled extrusion and deposition. This is done by using a paste of the material [18]. The FEF part building is conducted at a temperature below the freezing point of water, in order to solidify the paste while it is being deposited layer-after-layer to produce a green part that is then post-processed. This process has been used to produce ceramic components, such as alumina, zirconium diboride, and bioactive glass-scaffold parts [19–21].

The FEF is also used for the fabrication of FGM parts [22]. It uses a triple-extruder mechanism, each containing a paste of material. It uses a static mixer to blend the different material pastes into a homogeneous paste. The schematic diagram of the triple extruder is shown in Fig. 4.3. The green part is then freeze-dried at a below-freezing temperature and with a high pressure, which is held for several hours. After this stage, the part is then sintered at a high temperature—below the melting temperature of the constituent materials.

4.2.1 Functionally Graded Material Using the Material Extrusion Process

A number of research projects have been reported in the literature on the use of material extrusion for the fabrication of FGM [22–24]. Leu et al. [22] studied the

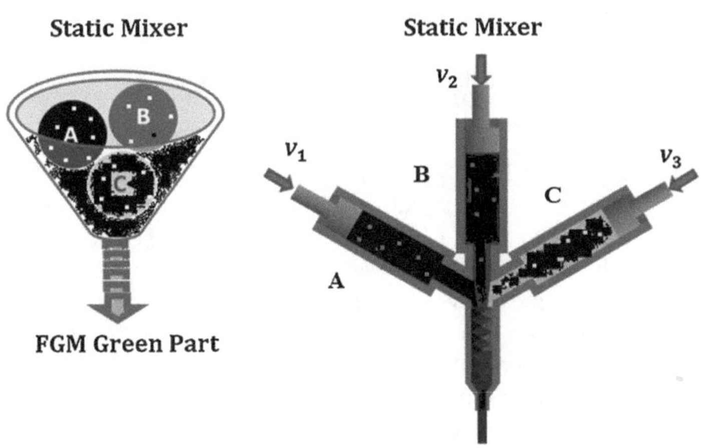

Fig. 4.3 Schematic diagram of static mixer and triple extruder

use of FEF for the production of FGM 3D objects. The study investigated the effectiveness of the developed FEF system through the fabrication of FGM of alumina (Al_2O_3) and zirconia (ZrO_2) parts. The green part was freeze-dried at a temperature of −25 °C and a pressure of 3000 Pa for 24 h. The samples were then allowed to warm up to room temperature in another 24 h. The samples were then sintered—by first heating the samples at 1 °C/min up to 600 °C; and the samples were then held for 1 h, in order to first burn out the organic binder. Thereafter, the samples were heated at 10 °C/min up to 1550 °C for another 90 min. The samples were then cooled back to room temperature at 25 °C/min. The sintered samples produced were analyzed by using energy-dispersive spectroscopy (EDS), in order to determine the FGM material composition. The results showed that there was a compositional change across the graded samples. This demonstrated the efficiency of using this technology for the fabrication of FGMs.

Srivastava et al. [23] investigated a technique to fabricate FGM, using the FDM technique. The FDM process parameters were mapped with the material properties for the given ABS material produced, and then used to develop the FGM samples. The study was able to establish that improved properties were achieved in the developed FGM. It was concluded that the work could be extended for modelling and simulating of the FDM components for different loading conditions.

The mechanical properties of FGM prototypes produced by FDM were also investigated by Li et al. [24]. Both the theoretical and the experimental analyses of the mechanical properties of the FGM prototypes produced were studied to develop constitutive models. The study resulted in a set of equations, which were developed for the determination of the elastic constants for the FDM prototypes. The developed models were validated through experimentation; and it was found that the model was in good agreement with the experimental data. The powder-bed fusion is presented in the next section.

4.3 Powder-Bed Fusion

The powder-bed fusion (PBF) process is a class of AM, where the powdered material is spread on the building platform (also known as the bed); the laser beam, or the electron beam, then scans the path generated by the 2D CAD profile from the sliced 3D CAD file. Each time a layer scanned is completed, the building platform is lowered by a layer thickness distance and the fresh powder is spread over the previously scanned layer; and the scanning process is repeated until the part-building process is completed. A hopper or a reservoir located below, or beside, the bed or building platform is used to provide the fresh powder material supply; while a roller or a blade is used to spread the powder over the building platform.

The schematic of the PBF system is shown in Fig. 4.4. The unspent powder (powder not scanned by the laser or electron) is used to support the object being built; and it is then removed from the part, when the building process is complete.

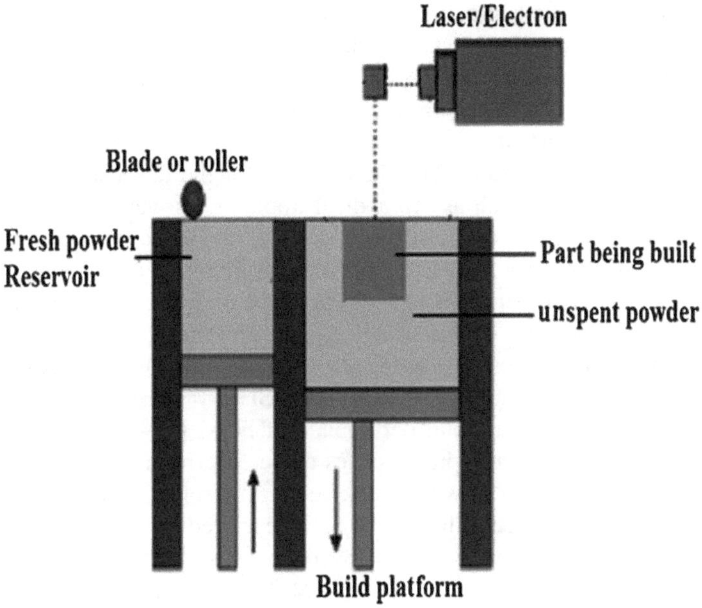

Fig. 4.4 Schematic diagram of PBF process

The AM technologies that belong to this class include the selective-laser sintering (SLS), the selective laser melting (SLM), the selective-heat sintering (SHS), and the electron-beam melting (EBM). The PBF uses a laser, or an electron beam, as the power source that is used to fuse or melt the powdered material together. The main difference between the SLS and the SLM process is that, the laser fuses the powder together in the SLS process; while the laser fully melts the powder in the SLM process. The SHS process is different from all other process; in that, the powder is fused together by a heated-thermal print head. The EBM process requires a vacuum for the creation of functional parts using metals and alloys; and it can also be used to fabricate FGM parts [25–28].

The areas of application of the PBF process include: prototypes, consumer products, architectural models, hardware, electronic housing, sculptures, and promotional items.

In the selective laser sintering process, the temperature of the building's platform chamber is controlled; and it is usually a bit below that of the material's melting temperature. The build chamber is also frequently filled with nitrogen gas, in order to protect the built part from environmental attack, and hence help to maintain the end quality and property of the object being built. One of the advantages of the SLS process is that the material efficiency is very high; because there is a possibility of recycling and reusing the unspent powdered materials. This was demonstrated in a study conducted by Ardila et al. [29] and Sayda et al. [30].

It was established that after the recycling of the powders between 12 and 14 times, there were no significant changes in the powder properties and the test parts' properties, which demonstrated that the high material efficiency in SLS was even higher than in SLM. The parts produced using the SLS are porous; and they are most suitable for the production of parts with the necessary porosity, such as the biomedical implants and filters. The SLS is used for the fabrication of prototypes, tooling, functional parts, such as biomedical implants, as well as parts made up of FGM [31–35]. The SLM is used to produce fully dense parts; and the process is faster than the SLS process. However, the process often requires the use of an inert gas in the building chamber. This is because of the high temperature involved, and the high level of reactivity of the processed part and the atmospheric oxygen.

The selective laser melting process is also an energy-intensive process and with a poor energy efficiency of about 10–20%. The SLM process is also used for the production of functional parts and parts made of FGM [36–39].

The selective heat sintering process uses a heated-print head to fuse the powder material together. Layers of powder are added with a roller; and the heated head is used to fuse the powder, according to the path described by the 2D CAD file sliced data. The platform is lowered by one layer thickness; and the powder is spread over the previous layer; and the fusing of the powder by the heated printing head is repeated. The process is repeated until the building is completed. The SHS process is used to produce prototypes, non-structural components and parts made of FGM [40–42].

The main advantage of this process is that the heat and power required by the thermal print head is significantly lower than those produced by the laser and the electron-beam processes.

The electrom beam melting is similar to the SLS and the SLM process, with the difference being that the electron beam is used as a power source in the EBM process, instead of the laser; and also, the EMB is used to fabricate metal parts only; whereas metals, ceramics, and composite materials can be used in the SLS and SLM processes. In the EBM process, the melting process is achieved by the use of a high-speed electron beam that strikes the surface of the metal powder. The kinetic energy of the striking electron beam and the metal powder are then converted into thermal energy, because of the reaction that takes place between the electron beam and the powder particle.

The thermal energy continues to increase; and it eventually rapidly melts the powder particles. The schematic diagram of the EBM process is shown in Fig. 4.5. Inside the electron-beam gun, a filament produces a cloud of electrons that stream out at a high speed. These fast-moving electrons hit on the powder molecules with high-impact energy that generates the high thermal energy. The EBM is an energy-efficient process with up to 95% efficiency; and it is also very fast. It is five to ten times faster than that of the laser-based AM technologies [43]. The EBM is used to fabricate end-used metallic parts and parts made of FGM [44–48].

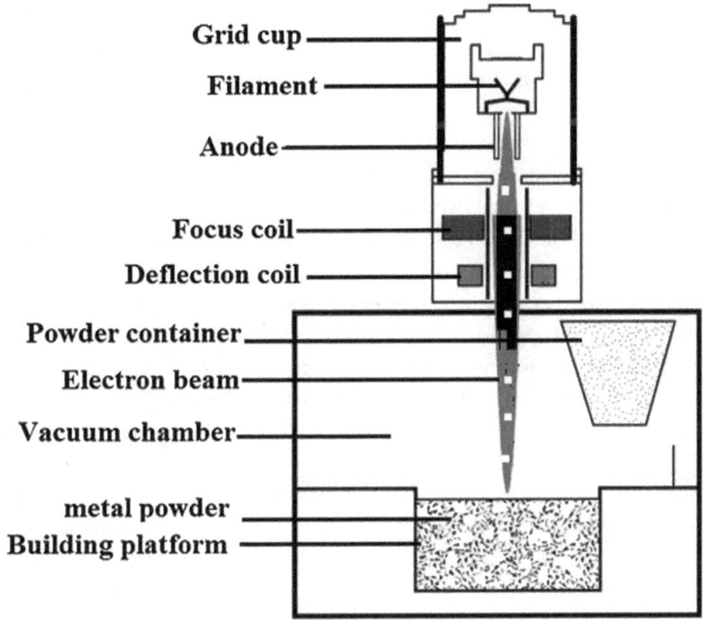

Fig. 4.5 Schematic diagram of the EBM process

4.3.1 Functionally Graded Material using the PBF Process

Much research work has been done on the SLS used for the production of FGM that has been reported in the literature [34, 35, 49, 50]. Haseung and Das [34] carried out studies on the FGM of Nylon-11 composites filled with different volume fractions of glass beads (0–30%). These were produced with the SLS process using theoretical modelling, numerical analysis, and experimental study. Each of the layers in the FGM were produced using the optimum-process parameters generated by the developed model, which was also experimentally validated. The properties of the developed FGM were investigated, and the result compared with the theoretical model, which was found to be in good agreement. The results were used to fabricate two components that exhibited a one-dimensional functionally graded material of particulate-filled polymer composites. The two different components that were designed were a compliant gripper and a rotator-cuff scaffold. The study concluded that the SLS is a powerful process for the fabrication of geometrically complex components made with FGM, using the appropriate powder-delivery systems. In a similar investigation by the same authors, a 3D structure made of FGM polymer Nano composite materials was produced and studied.

The properties of the FGM polymer Nano composites of Nylon-11 filled with 0–10 vol.% of 15 nm fumed silica nanoparticles were investigated. Optimum-process parameters were also developed for each of the layers in the FGM. The densities and micro/nanostructures of the Nano composites were studied; and the tensile property and the compressive properties for each composite were tested. The study revealed that there is a non-linear relationship between the processing parameters and the resulting properties in the one-dimensional polymer Nano composite FGM produced.

The result further confirmed the capability of using the SLS process for the production of one-dimensional compositional gradient 3D components with spatially varying mechanical properties.

Trainia et al. [49] investigated the properties of titanium alloy implants with gradient porosity fabricated by using the SLS process. The properties that were studied were: The surface roughness, the microstructure, the chemical composition, and the mechanical properties. The results revealed that the SLS process is an efficient process for producing dental implants with FGM that show better elastic properties very close to those of the human bone. The surface roughness was also found to match those of the human bone, which should enable osteo-integration to occur.

Sudarmadji et al. [50] studied the production of functionally graded scaffold using the selective-laser-sintering process. The mathematical relationship between the scaffold porosity and the compressive stiffness was established in this study. Different structural configurations and graded porosity scaffolds were produced. The toxicity of the fabricated scaffold was also accessed. The study revealed that the porosities, the compressive stiffness, and the yield strengths of the produced scaffolds were found to be closely matched that of the cancellous bone. The SLS has also been proven in this study to be an effective manufacturing process for the production of FGM implants.

Maskery et al. [39] studied the relationships between the lattice geometry and the mechanical behaviour of uniform and graded density $Al–Si_{10}–Mg$ lattices that were fabricated using the SLM process. The crushing behaviour of the FGM under the quasi-static loading was also studied. The results showed that the as-built sample showed a brittle collapse under the quasi-static loading condition. After the appropriate thermal treatments, the microstructure was seen to be changed and the energy-absorption property was greatly improved. The heat-treated graded lattices showed a progressive layer collapse with incremental strengthening.

The graded and the uniform structures absorbed almost the same amount of energy before the densification process; while the densification process was found to occur at around 7% lower strain for the graded structures. The study showed that the SLM, combined with the proper heat treatment, can be used to produce lightweight graded-lattice structures.

Hazlehurst et al. [51] investigated the production of a FGM femoral stem from a single alloy with varying mechanical properties using the SLM process. Finite-element analyses in the experimental study were used to investigate the three types of FGM with porosity-graded structures. This work investigated the flexural behaviour of FGM of cobalt–chrome femoral stems. The study showed that femoral stems with the FGM structure can be produced repeatedly with the SLM process—with a better flexural property and lighter weight. However, the study showed an inability to repeat the results in a sample that incorporated a cellular structure with a strut size of 0.5 mm.

Good agreement was obtained from the finite-element model and the experimental data, thereby showing that the simplified modelling approach used (the cellular structures were modelled as continuum parts and with a physically determined compressive elastic modulus) was suitable for modelling the behaviour of the femoral stems developed.

Tan et al. [28] investigated the properties of graded microstructure FGM of $Ti_6A_{l4}V$ produced using the EBM process. The microstructure consists of the columnar prior beta grains surrounded by a wavy grain boundary of alpha and transformed alpha/beta structures—with both the cellular colony and a basket-woven microstructure. The microstructures of the prior beta grains were found to be continuously increasing, as the build height was increased. This happened because the cooling rate reduces; as the build height increases. The reason for this is that at the beginning of the process, the substrate acts as heat sink; and its temperature increases as the build height is increased, which of course reduces the cooling rate. As a result of this, a graded microstructure was produced. The tensile property and the microhardness properties were also studied; and they were found to change across the volume of the samples. The microhardness, the yield strength, and the ultimate tensile strength increases with increases in build height. The elongation, on the other hand, was found to decrease; as the build height was increased.

Yan et al. [48] developed a multiscale heat-transfer modelling framework to investigate the EBM process for producing FGM. The developed heat source model describes the heating phenomenon that was based on a simulation of the micro-scale electron–material interactions. The model was used to predict the properties of the FGM produced. The developed model was found to be in good agreement with the experimental results. It was concluded that the theoretical model developed, could be used to predict the composition distribution of the FGM produced when using the EBM; and hence, it can be used in the design and manufacturing of FGM using the EBM process.

The selective heat sintering process was used in combination with a slip-casting for the manufacturing of $MoSi_2/Al_2O_3$ FGM component [52].

4.4 Directed-Energy Deposition

Directed-Energy Deposition (DED) is a class of AM process that is used to produce 3D objects from a 3D CAD model using the energy (such as laser, electron beam, or plasma arc) to create a melt pool on the substrate; while the powder or wire material, that is located coaxially with the energy source, is delivered into the melt pool. This would create a track of solid material upon solidification of the melt pool. The schematic diagram of the DED process is shown in Fig. 4.6. The process is repeated layer-after-layer to build a 3D solid part. The AM technologies that belong

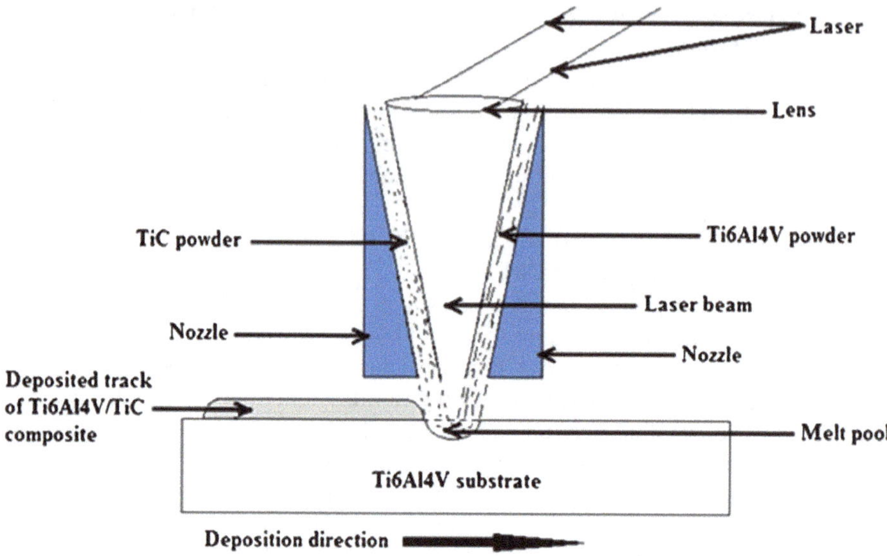

Fig. 4.6 Schematic diagram of DED process [57]

to these classes include: the laser-metal deposition process (LMD), also known as direct-metal deposition, or directed-light fabrication, which is what the Laser-engineered net shaping technology is based on, and electron-beam deposition. This is the only class of AM technologies that can be used for the repair of high valued components, which were not reparable in the past [53, 54].

This class of AM technology can also be used to build new material on an existing material base; and it is well suited in product remanufacturing. The DED can also be used to produce parts with FGM [55, 56].

In the directed energy deposition process, it is either the deposition head that is moved, while the object being built remains stationary on the build platform, in order to achieve the deposition of the 3D object; or it is the building platform that is moved, while the deposition head remain in a fixed position. Metals, alloys, and composite materials can be produced by using this process for the production of functional parts. The laser metal deposition process is an important additive manufacturing technology that belongs to the DED class of AM technologies. This technology uses a laser beam to create a melt pool on the substrate, in which wire or powder material is delivered and melted to create beads, tracks, and layers of solidified materials that are used to create the 3D solid object. The process is used to deposit functional parts made of metals, alloys, composite and FGMs [56–61]. Some of the advantages of the DED process include: the ability to control the grain structure by controlling the processing parameters. The ability to repair components and fabricate a new part on an existing part makes the process an attractive one and one with great potential. Some of the research work on LMD for the production of FGM is presented in the next subsection.

4.4.1 Functionally Graded Material Using the DED Process

The directed-energy deposition process has been successfully applied for the production of FGMs; and they have been reported in the literature [62–67]. Carroll et al. [62] conducted a thermodynamic study on the production of FGM of 304L stainless steel/Inconel 625 using the DED process to demonstrate the feasibility of producing FGN components when using the directed-energy deposition process. The microstructure (see Fig. 4.7), the chemical, the phase composition, and the microhardness properties of the FGM samples produced were investigated. The secondary-phase particles, consisting of transition metal-carbide particles were observed in their results. Thermodynamic computational modelling was used for the FGM. The model was validated by using an experimental study.

The result showed that the graded components can be produced using the directed-energy deposition process and the viability of the CALPHAD-based phase stability calculations was also demonstrated. The study concluded that there is a need to incorporate the possibility of secondary-phase particles; as the process may result in a variation of the properties of the FGM, depending on the build conditions employed.

Shah et al. [63] also conducted a similar research, using the laser-direct metal deposition (LDMD) process for the production of Steel 316L and Inconel 718 thin-wall FGM structures. The effect of process parameters, such as the laser power and the powder-flow rates on the microstructure, phase transformation, hardness, wear resistance, and tensile properties were studied.

The results showed that the continuously graded FGM can be produced by using the LDMD process; and that the properties of the FGM produced can be controlled by controlling the processing parameters.

The secondary dendritic arm spacing was observed in the microstructural study; and it was seen to be dependent on the powder mass-flow rate. The tensile strength of the FGM part was also found to be inversely proportional to the laser power; while the tensile strength was found to increase with an increase in the powder-flow rate.

Mahamood and Akinlabi [64] conducted research on the properties of FGM of Ti6Al4V/TiC produced using the LMD process. Two sets of samples were produced: one of the FGM sample was produced, using constant-process parameters; while the second sample was produced using the optimized process parameters obtained from a model that was previously developed for each of the material combinations in each level of the FGM. The mechanical and the tribological properties of the developed FGM were analyzed. The results showed that the FGM produced using the optimized process parameters has improved properties compared to those produced at constant processing parameters for all the material combinations in the FGM.

Balla et al. [65] studied the production of FGM of Ti–TiO_2, using the laser-engineered net shaping (LENS). The study showed that the deposited compositionally graded TiO_2 ceramic on porous Ti substrate significantly increased the surface wettability and hardness property of the substrate. The graded structures,

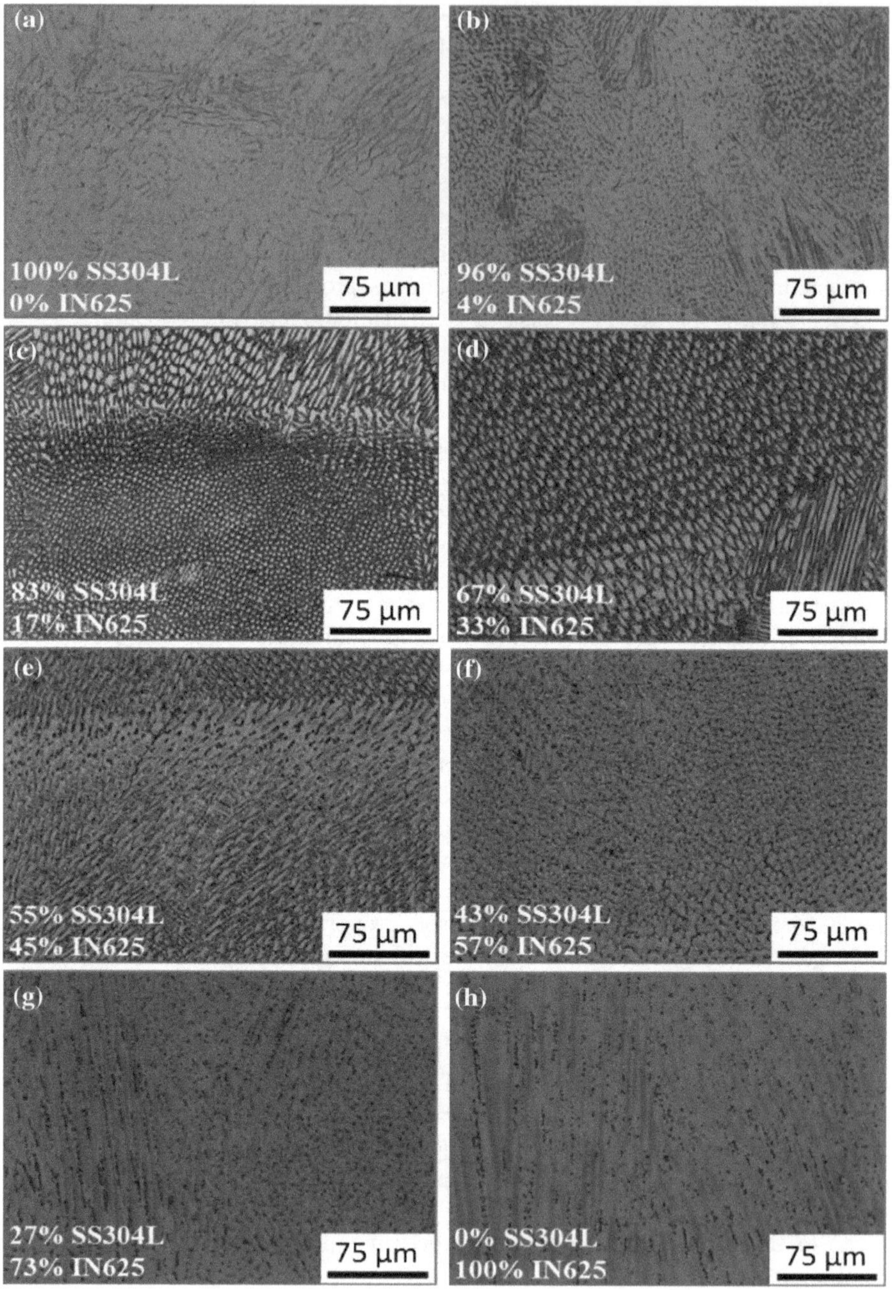

Fig. 4.7 Microstructure of gradient zone of the FGM of SS304L/IN625 (with permition from [62])

with the varying composition of TiO_2 on the top surface of the substrate, were found to be biocompatible. An improved wear property was also achieved. The study concluded that the FGM load-bearing implant with different compositions can be fabricated using the LENS process with an improved property.

The microhardness of the FGM produced was found to be approximately four times higher than that of the laser-processed pure Ti.

Durejko et al. [66] also conducted a study on the FGM of Fe–Al intermetallic alloys produced using LENS. A model was developed for the FGM, which was then verified through experimentation. The optimized process parameters showed that the FGM tubes produced have a good shape and a high metallurgical integrity. The FGM tubes have a smooth transition between both the components (the 316L steel and the Fe_3Al alloy), according to the chemical composition analysis and the microhardness results.

It was concluded that the high hardness of the two basic components achieved suggests that there is a need for heat treatment, in order to minimize the internal stresses produced in the FGM tubes produced.

4.5 Sheet Lamination

The sheet-lamination class of the AM process is a solid-state AM process. An example of this manufacturing process is the ultrasonic additive manufacturing (UAM) and the laminated object manufacturing (LOM). The laminated object manufacturing was invented by Helisys, Inc. of the USA in 1986 [67]. The schematic diagram of the sheet-lamination process is shown in Fig. 4.8. The LOM process uses adhesive-coated sheets that are bonded together with a heated roller that helps to glue the sheets together. The laser is then used to cut out the needed shape, according to the path from the CAD digital data. Although the original material that was used in this process was paper, which was basically used for the production of prototypes; other materials are now being used for the process, which include metals, ceramics, and composite materials [67, 68].

The development of new materials for the LOM process has made it possible to use the LOM process for the manufacturing of functional parts—apart from the

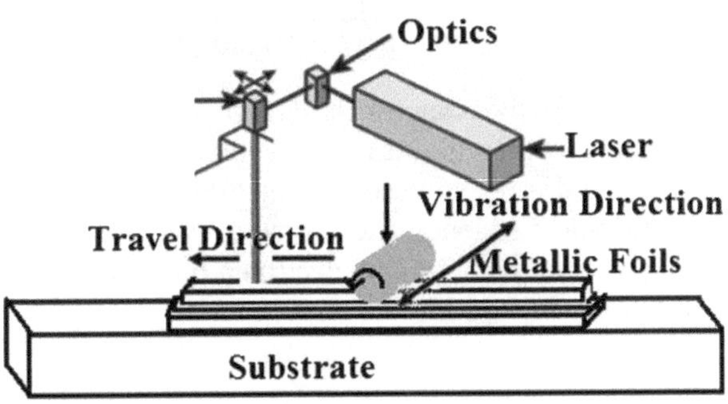

Fig. 4.8 Schematic diagram of ultrasonic consolidation

prototypes, for which the process was initially designed. The process does not require any support structure; because it is a self-supporting system for the overhangs and undercuts. The excess sheet material from the cut sheet material helps to form the needed support structure for the overhangs and the undercuts. There are other variants of this system, such as the Kira's paper lamination technology (PLT) [67, 69–71]. The LOM has also been reported for the fabrication of parts with functionally graded material [72]. The following steps are involved in the use of the LOM process to produce 3D objects:

- The sheet material is first adhered to the surface of the substrate with a heated roller. The heated roller helps to bind the sheet materials together.
- The laser beam is then used to trace the desired part, according to the CAD data information.
- The building platform is lowered; and the new sheet material is spread over the platform; and the process is then repeated.
- The laser cross-hatches on the areas that are not part of the process, in order to ease the waste material removal after the process is complete.

The main advantages of the LOM process are that the process is cheap; the process is fast; and the ease of the material handling in the process. The main disadvantage in this process is that the finishes achievable depend on the material used; and they may require post-processing to achieve the desired surface finish.

The Ultrasonic Additive Manufacturing (UAM) process is a similar process that uses sheets or ribbons of metal that are joined together by using the ultrasonic welding process. The UAM uses metallic materials, such as aluminium, copper, stainless steel, and titanium. The UAM process requires additional CNC machining process for the removal of unused materials during the welding process. The milling process can be performed after each layer is spread or added; or it can be performed after the entire bonding process. The UAM process is a low-temperature process that allows for the fabrication of parts with internal geometries.

The UAM has been used to produce metals, as well as FGMs [73, 74]. The process of the UAM involves the following steps:

- The material is positioned in place on the building platform;
- Adhesive material is used to bind the material in place over the previous layer;
- The required shape is traced out and then cut out from the building platform layer-by-layer using a laser, or a knife.
- The process is repeated until the building of the part is completed.

The main advantage of the UAM process is that the process can bind different materials together; and it requires relatively little energy; because it is a solid-state process; and it does not require melting of the materials; but it uses a combination of ultrasonic frequency and pressure. Another advantage of this process is that the overhangs can be built and electronic materials and wiring can be embedded in the 3D part produced [75].

4.5.1 *Functionally Graded Material Using Sheet-Ultrasonic Consolidation*

The use of a sheet-lamination class of AM technologies has also been reported, using the ultrasonic consolidation process [74]. Although there has not been much work reported in this area; Kumar [74] reported the production of FGM using the ultrasonic consolidation process. The FGM was produced by joining different metallic foils together. The schematic diagram of the cross-section of the FGM is shown in Fig. 4.9.

Three materials were used in this study; and they are: stainless steel, aluminium, and Copper foils. Optimum-process parameters for welding the combinations of the three materials was established and used to join the three materials together during the process. The microstructure and the microhardness of the produced FGM were studied. The graded structure is shown in Fig. 4.10 indicating the different compositions that made up the FGM. The functionally graded material could be produced by adopting the machining strategy, or using an intermediate glue layer. The study showed that the metallic FGM can be produced using the ultrasonic consolidation. Even though the process parameters that could be effectively varied in the process are limited.

Other research work on the production of FGM using the UAM process has also been reported in similar works of Domack and Baughman [76], Hu et al. [77] and White [78].

The laminated object manufacturing process has also been reported for the fabrication of FGMs [72, 79, 80]. The use of 3D printing and inkjet printing for the production of functionally graded materials has also been reported in the literature [81–85].

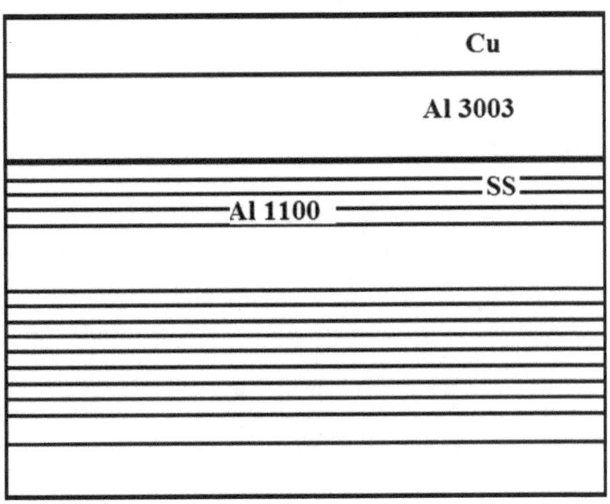

Fig. 4.9 Schematic diagram of FGM produced through the ultrasonic consolidation process

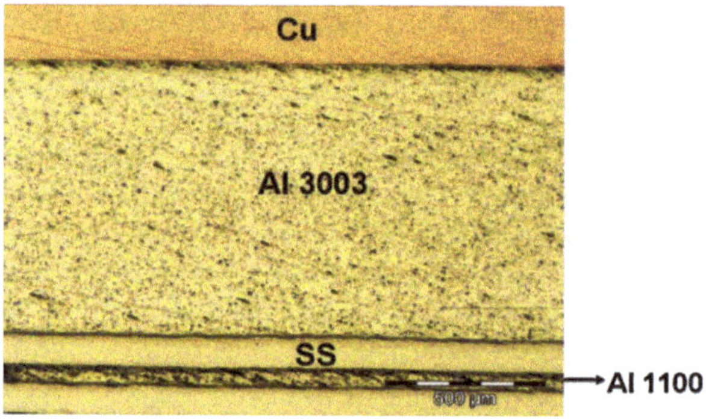

Fig. 4.10 The micrograph of the cross section of the functionally graded material with copper on the top and stainless steel below [74]

4.6 Summary

Additive manufacturing technology is a revolutionary manufacturing method that can produce 3D objects, no matter the complexity, directly from the 3D CAD model of the object by adding materials layer-by-layer. The different classes of AM technologies that can be used to fabricate FGM parts are presented in this chapter. They are: powder-bed fusion; material extrusion; sheet lamination and the powder-bed fusion process. The working principle of these additive manufacturing technologies was described; and some of the research projects in this area were also presented. The main advantage of using AM technologies for the production of FGMs is that FGM components can be produced, no matter the complexity—and in a single manufacturing process. The researchers showed that the improved properties in the FGM produced can be achieved by using these AM processes.

Acknowledgements This work was supported by the University of Johannesburg research council, the Department of Higher Education and Training (DHET) South Africa, the National Laser Centre Rental Pool Programme (RPP) contract number NLC - LREHA02-CON-001 and the L'Oreal-UNESCO for Women in Science.

References

1. Scott, J., Gupta, N., Wember, C., Newsom, S., Wohlers, T., Caffrey, T.: Additive manufacturing: status and opportunities, Science and Technology Policy Institute. Available from: https://www.ida.org/stpi/occasionalpapers/papers/AM3D_33012_Final.pdf (2012). Accessed on 11 Sept 2016
2. Mahamood, R.M., Akinlabi, E.T.: Achieving mass customization through additive manufacturing. In: Schlick, C., Trzcieliński, S. (eds.) *Advances in Ergonomics of Manufacturing: Managing the Enterprise of the Future*. Springer International Publishing Switzerland, pp. 385–390 (2016)
3. Mahamood, R.M., Akinlabi, E.T., Shukla, M., Pityana, S.: Evolutionary additive manufacturing: an overview. Lasers in Engineering **27**, 161–178 (2014)

4. Mahamood, R.M., Akinlabi, E.T., Shukla, M., Pityana, S.: Functionally graded material: an overview. In: Proceedings of the World Congress on Engineering, vol. III, pp. 1593–1597, WCE 2012, July 4–6, 2012, London, U.K. (2012)
5. Jacobs, P.F.: Rapid Prototyping & Manufacturing: Fundamentals of Stereolithography. SME Publication, Dearborn (1992)
6. Gibson, I., Rosen, D.W., Stucker, B.: Additive manufacturing Technologies, Rapid Prototyping to Direct Digital Manufacturing, 1st edn. Springer, New York (2010)
7. Sachs, M. E., Haggerty, J.S., Cima, M.J., Williams, P.A.: Three-Dimensional printing Techniques. US Patent, 5204055 (1993)
8. Comb, J.W., Priedeman, W.R., Turley, P.W.: FDM technology-process improvements. In: Proceedings of Solid Freeform Fabrication Symposium, pp. 42–49. Austin, TX
9. Beaman, J.J., Barlow, J.W., Bourell, D.L., Barlow, J.W., Crawford, R.H., McAlea, K.P.: Solid Freeform Fabrication: A New Direction in Manufacturing, pp. 25–49. Kluwer Academic Publishers, Norwell (1997)
10. Feygin, M., Hsieh, B.: Laminated Object Manufacturing (LOM): A Simpler Process. In: Proceedings of Solid Freeform Fabrication Symposium (pp. 123–130). Austin, TX (1991)
11. Mazumder, J., Schifferer, A., Choi, J.: Direct materials deposition: designed macro- and micro-structures. Mater. Res. Innovations **3**(3), 118–131 (1999)
12. Wohlers, T.: Additive manufacturing 101: Part I. Available at: http://www.wohlersassociates.com/JanFeb10TC.htm (2010). Accessed 3rd Oct 2016
13. Ahn, S.-H., Montero, M., Odell, D., Roundy, S., Wright, P.K.: Anisotropic material properties of fused-deposition modelling ABS. Rapid Prototyping **8**(4), 248–257 (2002)
14. Sood, A.K., Ohdar, R.K., Mahapatra, S.S.: Parametric appraisal of mechanical properties of fused-deposition modelling processed parts. Mater. Des. **31**, 287–295 (2010)
15. Es-Said, O.S., Foyos, J., Noorani, R., Mendelson, M., Marloth, R.: Effect of layer orientation on mechanical properties of rapid prototyped samples. Mater. Manuf. Processes **15**(1), 107–122 (2000)
16. Galantucci, L.M., Lavecchia, F., Percoco, G.: Study of compression properties of topologically optimized FDM-made structured parts. Manuf. Technol **57**, 243–246 (2008)
17. Huang, T., Mason, M.S., Hilmas, G.E., Leu, M.C.: Freeze-Form extrusion fabrication of ceramic parts. Int. J Virtual Phys Prototyping **1**(2), 93–100 (2006)
18. Mason, M.S., Huang, T., Landers, R.G., Leu, M.C., Hilmas, G.E.: Aqueous-based extrusion of high solids loading ceramic pastes: process modelling and control. J. Mater. Process. Technol. **209**(6), 2946–2957 (2009)
19. Huang, T., Mason, M.S., Hilmas, G.E., Leu, M.C.: Aqueous based freeze-form extrusion fabrication of alumina components. Rapid Prototyping J. **15**(2), 88–95 (2009)
20. Zhao, X., Landers, R.G., Leu, M.C.: Adaptive extrusion force control of freeze-form extrusion fabrication processes. ASME J. Manuf. Sci. Eng. **132**(6)
21. Doiphode, N.D., Huang, T., Leu, M.C., Rahaman, M.N., Day, D.E.: Freeze-Extrusion fabrication of 13-93 Bioactive glass scaffolds for bone repair. J. Mater. Sci. Mater. Med. **22**(3), 515–523 (2011)
22. Leu, M.C., Deuser, B.K., Tang, L., Landers, R.G., Hilmas, G.E. Watts, J.L.: Freeze-form extrusion fabrication of functionally graded materials. CIRP Ann. Manuf. Technol. **61**(1), 223–226
23. Srivastava, M., Maheshwari, S., Kundra, T.K.: Virtual modelling and simulation of functionally graded material components using FDM technique. Mater. Today Proc **2**(4–5), 3471–3480 (2015)
24. Li, L., Sun, Q., Bellehumeur, C., Gu, P.: Composite modelling and analysis for fabrication of FDM prototypes with locally controlled properties. J. Manuf. Processes **4**(2) (2002)
25. Wang, P., Tan, X., Nai, M.L.S., Tor, S.B., Wei, J.: Spatial and geometrical-based characterization of microstructure and microhardness for an electron-beam melted Ti–6Al–4V component. Mater. Des. **95**, 287–295 (2016)

26. Hinojos, A., Mireles, J., Reichardt, A., Frigola, P., Hosemann, P., Murr, l.E., Wicker, R.B.: Joining of Inconel 718 and 316 stainless steel using electron-beam melting additive-manufacturing technology. Mater. Des. **94**, 17–27
27. Tan, X., Kok, Y., Tan, Y.J., Vastola, G., Pei, Q.S., Zhang, G., Zhang, Y-W., Tor, S. B., Leong, K. F., Chua, C.K.: An experimental and simulation study on build thickness-dependent microstructure for electron-beam melted Ti–6Al–4V. J. Alloys Compd. **646**, 303–309 (2015)
28. Tan, X., Kok, Y., Tan, Y.J., Descoins, M., Mangelinck, D., Tor, S.B., Leong, K.F., Chua, C. K.: Graded microstructure and mechanical properties of additive manufactured Ti–6Al–4V via electron-beam melting. Acta Mater. **97**, 1–16 (2015)
29. Ardila, L.C., Garciandia, F., Gonzalez-Diaz, J.B., Alvarez, P., Echeverria, A., Petite, M.M., et al.: Effect of IN718 recycled powder reuse on properties of parts manufactured by means of selective laser melting. In: Physics Procedia of the 8th International Conference on Photonic Technologies LANE, vol. 56, pp. 99–107
30. Seyda, V., Kaufmann, N., Emmelmann, C.: (2012). Investigation of aging processes of Ti–6Al–4V powder material in laser melting. In: Physics Procedia of the 7th International Conference & Exhibition on Photonic Technologies LANE, vol. 39, pp. 425–31
31. Yan, M., Zhou, C., Tian, X., Peng, G., Cao, Y., Li, D.: Design and selective laser sintering of complex porous polyamide mould for pressure-slip casting. Mater. Des. **111**, 198–205 (2016)
32. Salmoria, G.V., Cardenuto, M.R., Roesler, C.R.M., Zepon, K.M., Kanis, L.A.: PCL/Ibuprofen implants fabricated by selective laser sintering for orbital repair. Procedia CIRP **49**, 188–192 (2016)
33. Du, Y., Liu, H., Shuang, J., Wang, J., Ma, J., Zhang, S.: Microsphere-based selective laser sintering for building macro-porous bone scaffolds with controlled microstructure and excellent biocompatibility. Colloids Surf. B **135**, 81–89 (2015)
34. Chung, H., Das, S.: Functionally graded Nylon-11/silica nanocomposites produced by selective laser sintering. Mater. Sci. Eng. A **487**(1–2), 251–257 (2008)
35. Chung, H., Das, S.: Processing and properties of glass bead particulate-filled functionally graded Nylon-11 composites produced by selective laser sintering. Mater. Sci. Eng. A **437**(2), 226–234 (2006)
36. Ahmadi, A., Mirzaeifar, R., Moghaddam, N.S., Turabi, A.S., Karaca, H.E., Elahinia, M.: Effect of manufacturing parameters on mechanical properties of 316L stainless steel parts fabricated by selective laser melting: a computational framework. Mater. Des. **112**, 328–338 (2016)
37. Pawlak, N., Rosienkiewicz, M., Chlebus, E.: Design of experiments approach in AZ_{31} powder selective laser-melting process optimization. Arch Civ. Mech. Eng. **17**, 9–18 (2017)
38. Matsumoto, R., Kanatani, S., Utsunomiya, H.: Filling of surface pores of aluminum foam with polyamide by selective laser melting for improvement in mechanical properties. J. Mater. Process. Technol. **237**, 402–408 (2016)
39. Maskery, I., Aboulkhair, N.T., Aremu, A.O., Tuck, C.J., Ashcroft, I.A., Wildman, R.D., Hague, R.J.M.: A mechanical property evaluation of graded density $Al–Si_{10}–Mg$ lattice structures manufactured by selective laser melting. Mater. Sci. Eng. A **670**, 264–274 (2016)
40. Sola, A., Bellucci, D., Cannillo, V.: Functionally graded materials for orthopedic applications —an update on design and manufacturing. Biotechnol. Adv. **34**(5), 504–531 (2016)
41. Gibson, I., Rosen, D.W., Stucker, B.: Additive Manufacturing Technology—Rapid Prototyping to Direct Digital Manufacturing. Springer, New York (2010)
42. Baumers, M., Tuck, C., Hague, R.: Selective heat sintering versus laser sintering: comparison of deposition rate, process energy consumption and cost performance. Available at: http://sffsymposium.engr.utexas.edu/sites/default/files/2015/2015-9-Baumers.pdf. Accessed online on 4th Oct 2016
43. Hiemenz, J.: Electron-beam melting. Adv. Mater. Processes, pp. 45–46 (2007)
44. Ruttert, B., Ramsperger, M., Roncery, L.M., Lopez-Galilea, I., Körner, C., Theisen, W.: Impact of hot isostatic pressing on microstructures of CMSX-4 Ni-base superalloy fabricated by selective electron-beam melting. Mater. Des. **110**, 720–727 (2016)

45. Fiaz, Hasan S., Settle, Casey R., Hoshino, Kazunori: Metal-additive manufacturing for micro-electro-mechanical systems: Titanium alloy (Ti–6Al–4V)-based nanopositioning flexure fabricated by electron-beam melting. Sens. Actuators A **249**(1), 284–293 (2016)
46. Algardh, J.K., Horn, T., West, H., Aman, R., Snis, A. Engqvist, H., Lausmaa, J., Harrysson, O.: Thickness dependency of mechanical properties for thin-walled titanium parts manufactured by Electron-Beam Melting (EBM)®. Addit. Manuf. **12**, Part A, 45–50
47. Portolés, L., Jordá, O., Jordá, L., Uriondo, A., Esperon-Miguez, M., Perinpanayagam, S.: A qualification procedure to manufacture and repair aerospace parts with electron-beam melting. J. Manuf. Syst. **41**, 65–75 (2016)
48. Yan, W., Ge, W., Smith, J., Lin, S., Kafka, O.L., Lin, F., Liu, W.K.: Multi-scale modelling of electron-beam melting of functionally graded materials. Acta Mater. **115**, 403–412 (2016)
49. Trainia, T., Mangano, 1, C., Sammons, R.L., Mangano, F., Macchib, A., Piattelli, A.: Direct laser-metal sintering as a new approach to fabrication of an isoelastic functionally graded material for the manufacture of porous titanium dental implants. Dent. Mater. **24**, 1525–1533 (2008)
50. Sudarmadji, N., Tan, J.Y., Leong, K.F., Chua, C.K., Loh, Y.T.: Investigation of the mechanical properties and porosity relationships in selective laser-sintered polyhedrals for functionally graded scaffolds. Acta Biomater. **7**, 530–537 (2011)
51. Hazlehurst, K.B., Wang, C.J., Stanford, M.: An investigation into the flexural characteristics of functionally graded cobalt-chrome femoral stems manufactured using selective-laser melting. Mater. Des. **60**, 177–183 (2014)
52. Dumont, A-L., Bonnet, J-P., Chartier, T., Ferreira, J.M.F.: $MoSi_2/Al_2O_3$ FGM: elaboration by tape casting and SHS. J. Eur. Ceram. Soc. **21**, 2353–2360 (2001)
53. Pinkerton, A.J., Wang, W., Li, L.: Component repair using laser direct-metal deposition. J. Eng. Manuf. **222**, 827–836 (2008)
54. Graf, B., Gumenyuk, A., Rethmeier, M.: Laser-metal deposition as repair technology for stainless steel and Titanium alloys. Phys. Procedia **39**, 376–381 (2012)
55. Mahamood, R.M., Akinlabi, E.T., Shukla M., Pityana, S.: Functionally graded material: an overview. Proc. World Congr. Eng. WCE **3**, 1593–1597 (2012)
56. Mahamood, R.M., Akinlabi, E.T.: Laser-metal deposition of functionally graded Ti6Al4V/TiC. Mater. Des. **84**, 402–410 (2015)
57. Mahamood, R.M., Akinlabi, E.T., Shukla, M., Pityana, S.: scanning velocity influence on microstructure, microhardness and wear-resistance performance on laser-deposited $Ti_6A_{14}V$/TiC composite. Mater. Des. **50**, 656–666 (2013)
58. Mahamood, R.M., Akinlabi, E.T., Shukla, M., Pityana, S.: Characterizing the effect of laser-power density on microstructure, microhardness and surface finish of laser-deposited titanium alloy. J. Manuf. Sci. Eng. **135**(6), doi:10.1115/1.4025737 (2013)
59. Mahamood, R.M., Akinlabi, E.T., Shukla, M., Pityana, S.: Effect of laser power on material efficiency, layer height and width of laser-metal deposited $Ti_6A_{14}V$, pp. 1433–1438. World Congress of Engineering and Computer Science, San Francisco (2012)
60. Mahamood, R.M., Akinlabi, E.T., Shukla, M., Pityana, S.: Characterization of laser-deposited Ti_6A_4V/TiC composite. Lasers Eng. **29**(3–4), 197–213 (2014)
61. Milewski, J.O., Lewis, G.K., Thoma, D.J., Keel, G.I., Nemec, R.B., Reinert, R.A.: Directed-light fabrication of a solid metal hemisphere using 5-axis powder deposition. J. Mater. Process. Technol. **75**, 165–172 (1998)
62. Carroll, B.E., Otis, R.A., Borgonia, J.P., Suh, J.-O., Dillon, R.P., Shapiro, A.A., Hofmann, D. C., Liu, Z.-K., Beese, A.M.: Functionally graded material of 304L stainless steel and Inconel 625 fabricated by directed-energy deposition: characterization and thermodynamic modelling. Acta Mater. **108**, 46–54 (2016)
63. Shah, K., Haq, I.U., Khan, A., Shah, S.A., Khan, M., Pinkerton, A.J.: Parametric study of development of inconel-steel functionally graded materials by laser direct-metal deposition. Mater. Des. **54**, 531–538 (2014)
64. Mahamood, R.M., Akinlabi, E.T.: Laser-metal deposition of functionally graded $Ti_6A_{14}V$/TiC. Mater. Des. **84**(5), 402–410 (2015)

65. Balla, V.K., DeVasConCellos, P.D., Xue, W., Bose, S., Bandyopadhyay, A.: Fabrication of compositionally and structurally graded Ti–TiO_2 structures using laser-engineered net shaping (LENS). Acta Biomater. **5**, 1831–1837 (2009)
66. Durejko, T., Ziętala, M., Polkowski, W., Czujko, T.: Thin-wall tubes with $Fe_3Al/SS_{316}L$ graded structure obtained by using laser-engineered net-shaping technology. Mater. Des. **63**, 766–774 (2014)
67. Boboulos, M.A.: CAD-CAM & rapid prototyping application evaluation. Retrieved 21st September 2016 from www.bookBoom.com (2011)
68. Klosterman, D.A., Chartoff, R.P., Osborne, N.R., Graves, G.A., Lightman, A., Han, G., Bezeredi, A., Rodrigues, S.: Curved layer LOM of ceramics and composites. In: Proceedings of the 1998 Solid Freeform Fabrication Symposium, pp. 671–680 (1998)
69. Banerjee, S.: Development of a novel toner for electrophotography-based additive manufacturing process. Ph.D. thesis, De Montlbrt University (2009)
70. Nakagawa, T.: Recent developments in auto-body panel forming technology. CIRP Ann. Manuf. Technol. **42**(2), 1\1–121 (1993)
71. Nakagawa, T., Kunieda, M.: Manufacturing of laminated deep-drawing dies by laser-beam cutting. Adv. Technol. Plast. **1**, 520–525 (1984)
72. Zhang, Y., Han, J., Zhang, X., He, X., Li, Z., Du, S.: Rapid prototyping and combustion synthesis of TiC/Ni functionally gradient materials. Mater. Sci. Eng. A **299**, 218–224 (2001)
73. White, D.R.: Ultrasonic consolidation of aluminium tooling. Adv. Mater. Processes 64–65 (2003)
74. Kumar, S.: Development of functionally graded materials by ultrasonic consolidation. CIRP J. Manuf. Sci. Technol. **3**, 85–87 (2010)
75. Kong, C.Y., Soar, R.C., Dickens, P.M.: Ultrasonic consolidation for embedding SMA fibres within aluminium matrices. Compos. Struct. **66**, 421–427 (2005)
76. Domack, M.S., Baughman, J.M.: Development of nickel-titanium-graded composition components. Rapid Prototyping J. **11**(1), 41–51 (2005)
77. Hu, Y., Fadel, G.M., Blouin, V.Y., White, D.R.: Optimal design for additive manufacturing of heterogeneous objects using ultrasonic consolidation. Virtual Phys. Prototyping **1**(1), 53–62 (2006)
78. White, D.R.: Ultrasonic consolidation of aluminium tooling. Adv. Mater. Processes **161**, 64–65 (2003)
79. Tari, M.J., Bals, A., Park, J., Lin, M.Y., Thomas, Hahn H.: Rapid prototyping of composite parts using resin-transfer moulding and laminated-object manufacturing. Compos. Part Appl. Sci. Manuf. **29**(5–6), 651–661 (1998)
80. Klosterman, D., Chartoff, R., Graves, G., Osborne, N., Priore, B.: Interfacial characteristics of composites fabricated by laminated object manufacturing. Compos. Part. Appl. Sci. Manuf. **29**(9–10), 1165–1174 (1998)
81. Jackson, T.R., Liu, H., Patrikalakis, N.M., et al.: Modelling and designing functionally graded material components for fabrication with local composition control. Mater. Des. **20**(2–3), 63–75 (1999)
82. Dimitrov, D., Schreve, K., De Beer, N.: Advances in three-dimensional printing—state of the art and future perspectives. Rapid Prototyping J. **12**(3), 136–147 (2006)
83. Chartoff, R., McMorrow, B., Lucas, P.: Functionally Graded Polymer-Matrix Nano-Composites by Solid Freeform Fabrication: A Preliminary Report. In: Solid Freeform Fabrication Symposium Proceedings, Austin, Texas, pp. 385–391
84. Shevchenko, A.V., Dudnik, E.V., Ruban, A.K., et al.: Functional graded materials based on ZrO_2 and Al_2O_3 production methods. Powder Metall. Met. Ceram. **42**(3–4), 145–153 (2003)
85. Wang, J., Shaw, L.L.: Fabrication of functionally graded materials via inkjet colour printing. J. Am. Ceram. Soc. **89**(10), 3285–3289 (2006)

Chapter 5
Experimental Analysis of Functionally Graded Materials Using Laser Metal Deposition Process (Case Study)

5.1 Introduction

Functionally graded materials (FGMs) have evolved from the initial thermal-barrier application, for which it was developed. The FGM is now used for other applications, such as bullet-proof vests for penetration resistance application, applications that require a better wear-resistant property and applications, where high-performance composite materials are desirable. Different types of FGM have been developed for various applications; and these were presented in Chapter Two. The functionally graded titanium-alloy composite is one of these that has found its use in a number of applications.

Titanium and its alloys are important engineering materials, because of their impressive properties, such as high corrosion resistance and high strength-to-weight ratio [1, 2]. Titanium alloy, Ti64, is the most widely used titanium alloy, because of the exciting properties it possesses. It is referred to as the workhorse in the titanium industry; because it is the material with the highest strength-to-weight ratio [3]. Despite all these important properties, the wear-resistant property of these materials is poor, because of their low thermal conductivity and the high coefficient of friction of this material [4, 5]. Titanium and its alloys tend to be attracted to other materials with which they come into contact. This is the main reason why titanium and its alloys are termed as being difficult materials to machine [6]. The titanium reacts with the cutting tool material, thereby generating high temperatures that cause the titanium to adhere to the surface of the cutting tool; and eventually it breaks. This is why titanium needs surface modification; if it is to be used where it would come in contact with other materials. In an extreme working environment, the conventional titanium-alloy composite will fail in such working condition, hence the need for a FGM titanium-alloy composite. The difficulty experienced during the manufacturing of materials, such as the titanium alloy demands an advanced manufacturing process, such as the additive-manufacturing process.

R.M. Mahamood and E. Titilayo Akinlabi, *Functionally Graded Materials*, Topics in Mining, Metallurgy and Materials Engineering, DOI 10.1007/978-3-319-53756-6_5

The laser metal deposition (LMD) process, an advanced manufacturing process, belongs to the directed energy deposition (DED) class of additive manufacturing, which was dealt with extensively in Chapter Four of this book. The LMD process is an excellent alternative manufacturing process for processing titanium and its alloys. The LMD process is a laser-based additive-manufacturing process that can be used to produce three-dimensional (3D) parts directly from the 3D computer-aided design (CAD) model of the part being made—by adding materials in a layer-wise manner [7].

The main advantage of this LMD process is in its ability to make composite materials, as well as functionally graded materials, because of its capability to handle more than one material at the same time. The LMD process can be used produce a complex part; while at the same time, together with the FGM, in one manufacturing step. Considerable research work has been reported in the literature on the use of LMDs for the production of composite materials [8–11]. Some of these works include the study conducted by Obiolodan and Strucker [8]. The LMD process was used to produce composite materials of 10w% TiC/Ti6Al4V and 5w% TiC/Ti6Al4V on Ti6Al4V substrate.

Popoola et al. [9] studied the fabrication of TiC/Ti6Al4V composite at various TiC compositions using the LMD. Wang et al. [10] also conducted a similar research project by depositing a TiC/Ti6Al4V composite at different TiC compositions with the help of the LMD process. A similar study was also conducted, using the LMD process to produce a Ti/TiC composite, by Ochonogor et al. [11]. In that study, the influence of the TiC ratio on the wear-resistance property of the Ti/TiC composite was investigated.

Considerable research work has also appeared in the literature on the use of LMD for producing functionally graded materials. Some of the focus of the earlier works was aimed at establishing the feasibility of using LMD for the production of FGM [12–14]. Zang et al. [15] investigated the properties of FGM of Ti/TiC on a Ti6Al4V substrate, using the LMD process. The processing parameters for the various volume fractions of Ti/TiC composite were first established in their preliminary work. The results from this preliminary work were used to deposit a thin wall FGM of Ti/TiC using different processing parameters for each level of the FGM, during the deposition process. The result showed that the wear-resistance performance of the Ti6Al4V substrate was improved with the addition of the TiC. It was also shown that FGM can be produced by using the LMD process with a continuously changing microstructure. In another study performed by Wang et al. [16], the property of FGM of Ti6Al4V/TiC produced using the LMD process was investigated. Ti6Al4V wire and TiC powder were used for the production of the FGM; and the two materials were fed simultaneously into the melt-pool. The wire-feed rate of the Ti6Al4V was kept constant; while the TiC powder feed rate was varied. Liu and DuPont [17] also deposited FGM of Ti/TiC composite, using the LMD process. Mahamood and Akinlabi [18], in another study, investigated the wear-resistance property and the microhardness properties of FGMs of Ti6Al4V/TiC composite when using the LMD process. Optimized process parameters were used for the production of each compositional ratio of each layer of the FGM produced.

The optimized process parameters were obtained from a model that was earlier developed by the authors. The properties of the FGM produced were studied; and the result showed that the wear-resistance property and the microhardness properties were improved, when compared with the substrate material and the FGM produced at constant process parameters. The studies that were reported on the use of the LMD process for the production of composite materials, as well as the FGM included in the research work reported by Shah 2011 [19], Lin et al. [20], Qin et al. [21], and Schwendner et al. [22].

The case study that is presented in this chapter is the production of FGM of Titanium alloy Ti64 and TiC composite, when using the LMD process. Eleven layers of varying composition of the Ti64/TiC mixture were deposited to produce the FGM. The percentage content of each layer of the FGM was varied from 0% TiC to 50% TiC. The FGM produced was analyzed and the properties produced were studied and they were compared with the traditional composite material of Ti64/TiC [23] and the parent material (PM) of Ti64. The results are presented and explained in the following sections.

5.2 Materials and Methods

TiC powders with the purity of 99.5%, the gas-atomized Ti64 powders with the purity of 99.6% and the Ti64 square plate of 6 mm thickness with a purity of 99.6% were used in this case study. The particle size range of the TiC powder used was below 60 μm, while that of the Ti64 powder was between 120 and 350 μm. The particle size analysis is shown in Fig. 5.1. The Ti64 substrate was in an annealed condition; and it was supplied by VSMPO-AVISMA Corporation Company, Russia. The TiC and the Ti64 powder used in this case study were supplied by F. J. Brodmann and Co., L.L.C., Louisiana. The chemical composition of the TiC powder, the Ti64 powder, and the Ti64 substrate are presented in Tables 5.1, 5.2, and 5.3, respectively. The morphology of the TiC and the Ti64 powders is shown in Fig. 5.2a, b, respectively.

The Ti64 is in the form of gas-atomized powders with a spherical shape and smooth surfaces. Spherically shaped powders exhibit low surface oxidation, because of their reduced total surface area. The TiC powder is a ball-milled powder with irregular shapes, which is typically characteristic of a ball-milled powder. Before the deposition experiment, the Ti64 substrate was sandblasted to roughen the surface of the substrate and cleaned with acetone to degrease it. These steps are done to help to improve the laser-absorption process. The laser metal deposition process of the FGM composite was carried out using a Kuka robot equipped with a

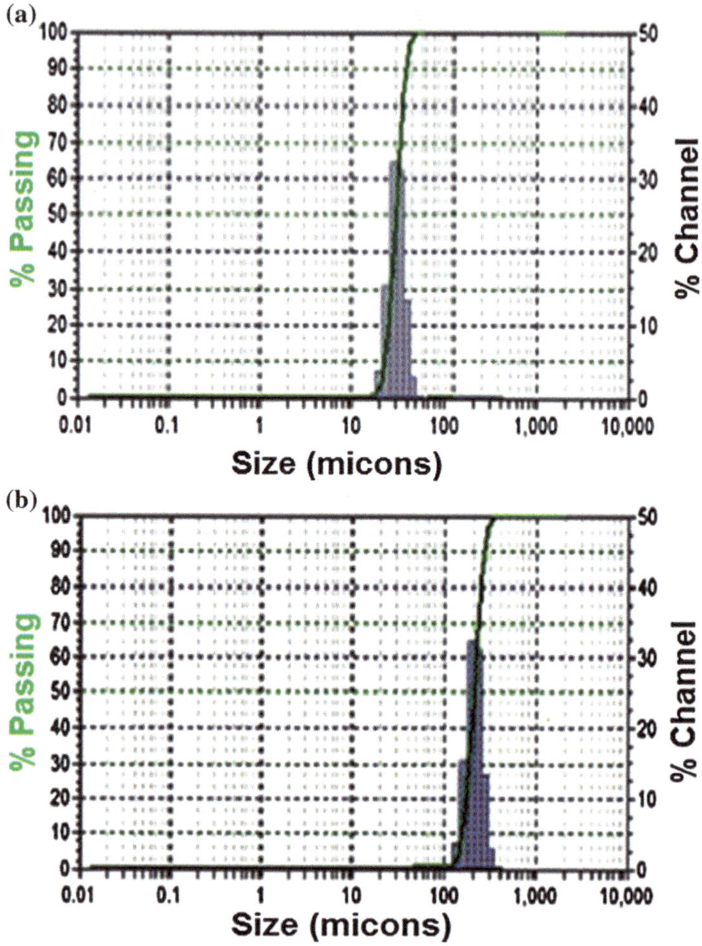

Fig. 5.1 Particle size analysis of **a** TiC powder, **b** Ti64 powder

Table 5.1 Chemical composition of the TiC powder

Element	C	O_2	N_2	Fe	Al	Na	Ti
W%	19.5	0.28	0.4	0.045	0.03	0.026	Balance

Table 5.2 Chemical composition of the Ti6Al4V powder

Element	Al	V	Fe	C	N_2	H_2	O_2	Ti
W%	6.20	3.90	0.18	0.008	0.005	0.005	0.150	Balance

Table 5.3 Chemical composition of Ti6Al4V substrate

Element	Al	V	Fe	C	N_2	H_2	O_2	Ti
W%	6.42	3.91	0.19	0.008	0.006	0.004	0.155	Balance

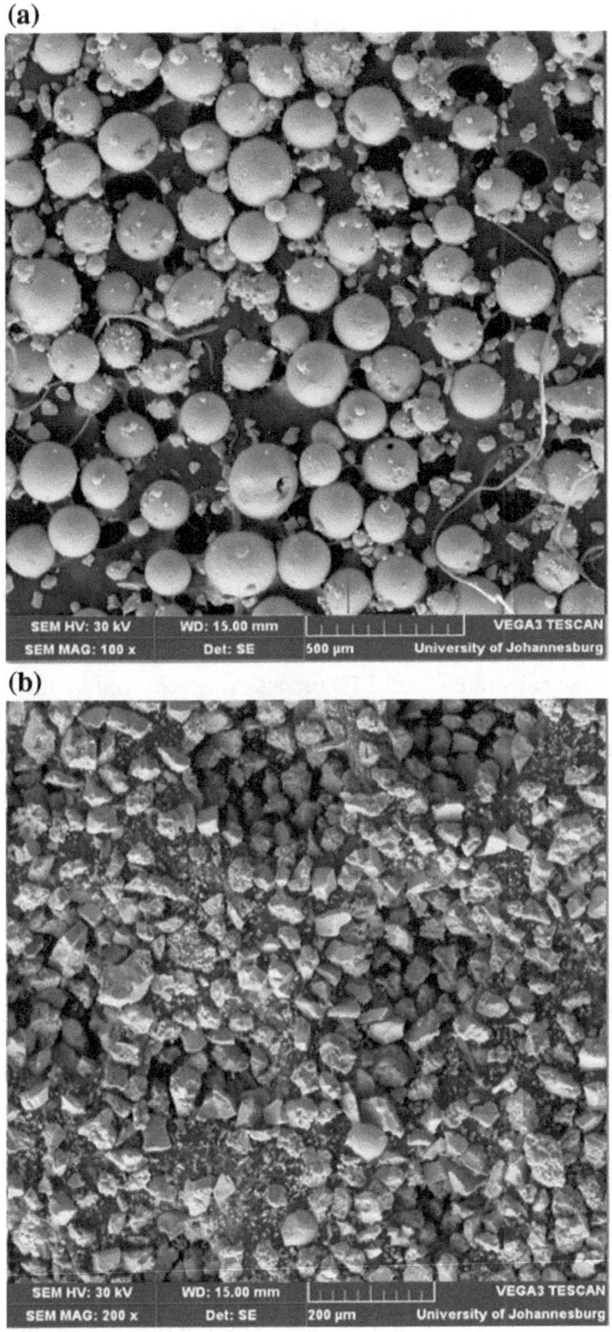

Fig. 5.2 The morphology of the **a** Ti64 powder, **b** TiC powder

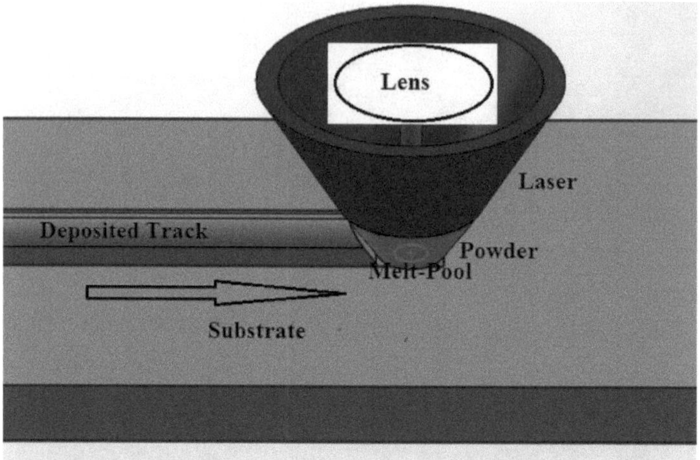

Fig. 5.3 Schematic diagram of the LMD process

4 kW Nd-YAG fibre laser that is available at the National Laser Centre (NLC) in the Centre for Scientific and Industrial Research (CSIR) in Pretoria, South Africa.

The diameter of the laser beam was maintained at 2 mm at a focal distance of 195 mm above the substrate. The LMD process was achieved by the laser creating a melt-pool on the surface of the substrate; while the TiC powder and the Ti64 powder were deposited into the melt-pool through the co-axial nozzles simultaneously.

The schematic diagram of the LMD process and the experimental set-up are shown in Figs. 5.3 and 5.4, respectively. The processing parameters that are used for the LMD experiment are presented in Table 5.4, according to Mahamood and Akinlabi [18].

Eleven layers of varying compositional ratio of Ti64 and TiC powder were deposited each at the optimized process parameters presented in Table 5.4. The first layer is 100% Ti64; while the subsequent levels consisted of less than 5% Ti64 and plus 5% TiC—with the last layer being composed of 50% of Ti64 and 50% of TiC. For comparison, a conventional composite Ti64/TiC was also deposited that consisted of 50% Ti64 and 50% of TiC.

The processing parameter used for the 50% Ti64 in Table 5.4 was used to deposit the conventional composite.

After the LMD process, the specimens for metallographic observation were prepared by cutting the samples through the transverse direction across the deposition direction. The PM, that is, the Ti64 plate, was also cut. The cut samples were mounted, ground, polished, and etched, according to the standard metallurgical preparation of titanium and its alloys [24]. The microstructures were studied by

Fig. 5.4 Pictorial diagram of the LMD experimental set-up

Table 5.4 Process parameters of FGM [17]

Layer designation	% TiC	Laser power (kW)	Scanning speed (m/s)	Powder flow rate (rpm)	Gas flow rate (l/min)
A	0	2.0	0.01	2.00	2.00
B	5	2.0	0.0075	3.00	2.00
C	10	2.2	0.0075	2.00	2.00
D	15	2.26	0.0075	2.00	2.00
E	20	2.21	0.0077	2.00	2.00
F	25	2.19	0.0078	2.00	2.00
G	30	2.16	0.0079	2.00	2.00
H	35	2.14	0.0080	2.00	2.00
I	40	2.12	0.0081	2.00	2.00
J	45	2.10	0.0082	2.00	2.00
K	50	2.07	0.0083	2.00	2.00

using the Olympus Optical microscope by BX51M; Olympus and the Scanning Electron Microscope (SEM) by TESCAN equipped with the Oxford Energy Dispersion Spectrometry (EDS).

The polished samples were tested for their microhardness, using the microhardness tester from Metkon. The microhardness profiling was taken with a load of 500 g and a dwelling time of 15 s. The distance between the indentations was maintained at 12 μm. A minimum of inter-indentation distance of more than twice the indentation diameters of the indentation was prescribed by the ASTM standard on microhardness measurement [25]. The wear-resistance properties of all the samples were studied on the deposited samples that had been ground to 1000 μm, using the ball-on-disk tribotester produced by Cert with a tungsten carbide ball of 10 mm diameter.

A load of 25 N, a sliding distance of 2000 mm, and the sliding speed of 0.02 m/s were the experimental conditions were used during the wear test. The tests were carried out in the dry air without any lubrication. The sliding wear was carried out, according to the ASTM standard on wear measurement [26]. The wear scars were studied under the SEM and the wear volumes were calculated by using the equation developed by Sharma et al. [27].

The phases present were studied by using the X-ray diffraction analysis by Ultima IV.

5.3 Results and Discussion

The microstructure of the Ti64 is shown in Fig. 5.5. The microstructure of the substrate is made up of the beta phase in the matrix of alpha phase particles. The alpha phase comprises the white parts shown in the diagram. The dark parts are the beta phase, which is the characteristic of a typical Ti64, titanium alloy. The micrograph of the FGM is shown in Fig. 5.6. Each level of the FGM composition is shown in the diagram. The microstructure showed a continuous structure—without any distinct interface. It can be seen that the microstructure changes with the height of the FGM. This is as a result of the changing percentage composition of the TiC. This has also demonstrated that the LMD process can be used to produce a continuous microstructure for the FGM.

The graph of the microhardness against the test distance of the FGM sample, the composite Ti6a/TiC and the Ti64 PM is shown in Fig. 5.7. The micrograph of the FGM showing the microhardness indentation is shown in Fig. 5.8. The bar chart of the average microhardness is shown in Fig. 5.9.

It can be seen from Figs. 5.7 and 5.9 that the FGM has the maximum average microhardness values. From the Fig. 5.7, the microhardness value of the topmost part of the FGM sample is seen to be as high as 1200 HV. This is as high as four

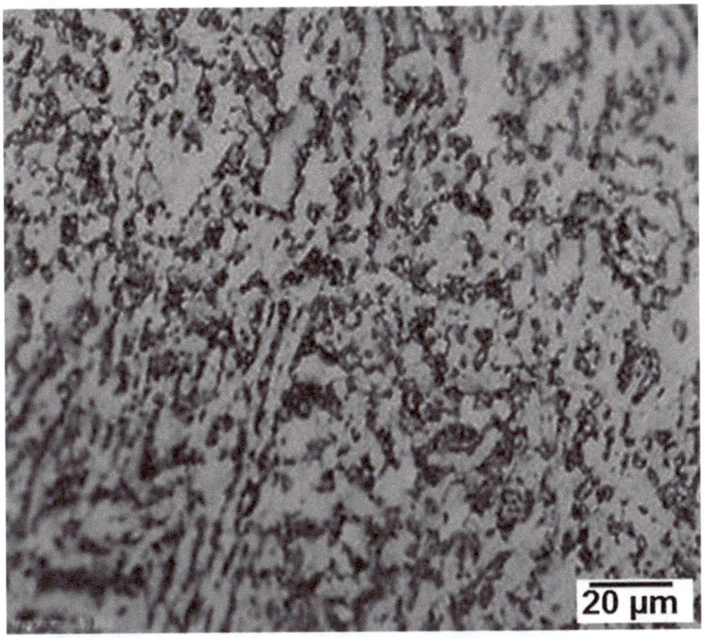

Fig. 5.5 The micrograph of Ti64 substrate

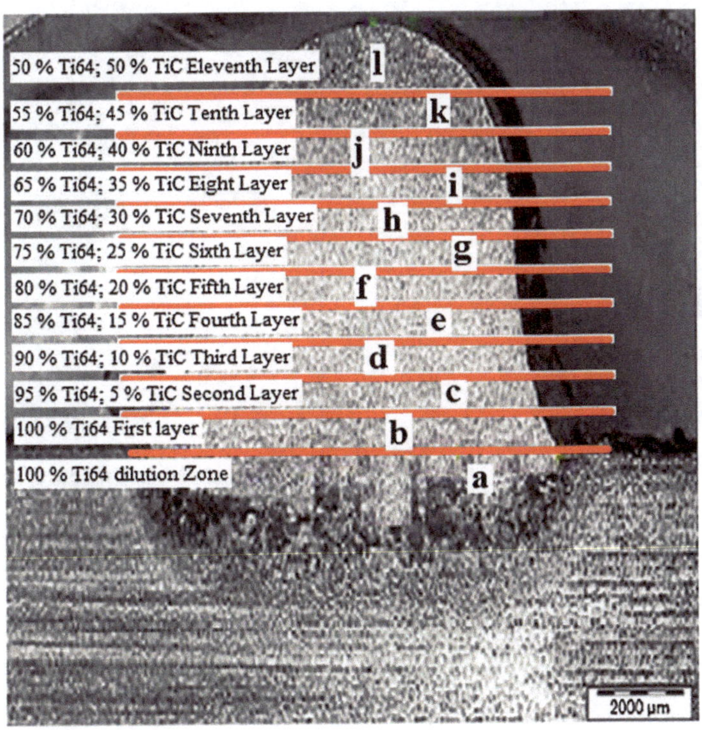

Fig. 5.6 The micrograph of FGM of the Ti64/TiC

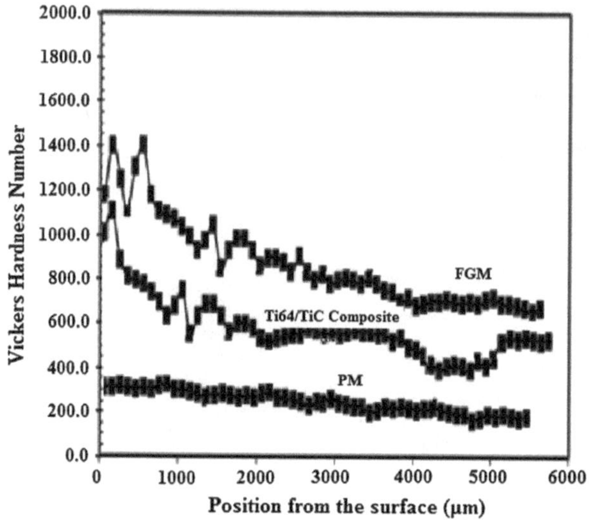

Fig. 5.7 The graph of the microhardness against the distance from the surface

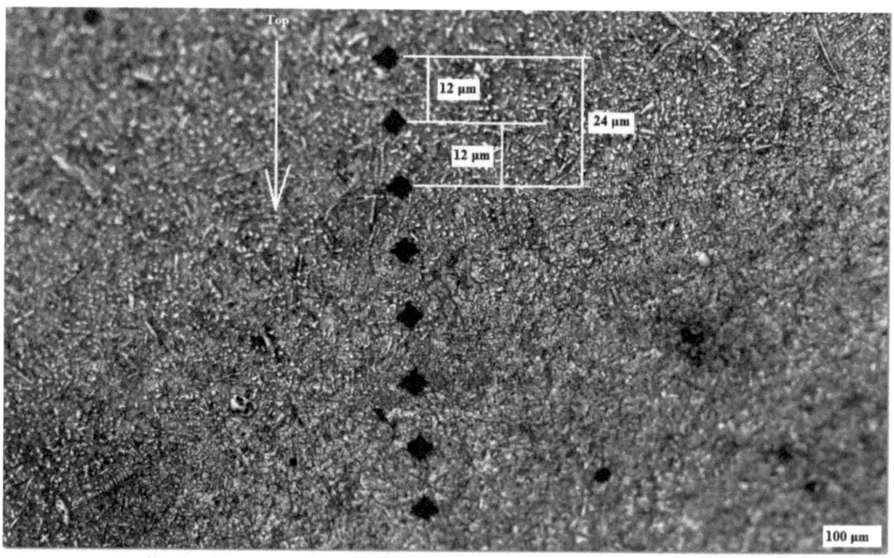

Fig. 5.8 The optical micrograph showing the microhardness indentation on the FGM sample

times the microhardness value of the substrate. The high hardness value seen is as a result of the TiC content, and because of the dendritic TiC seen in the microstructure.

The wear results of the three samples are presented in Table 5.5 and the bar chart of the wear volume of the FGM, the composite and the PM are presented in

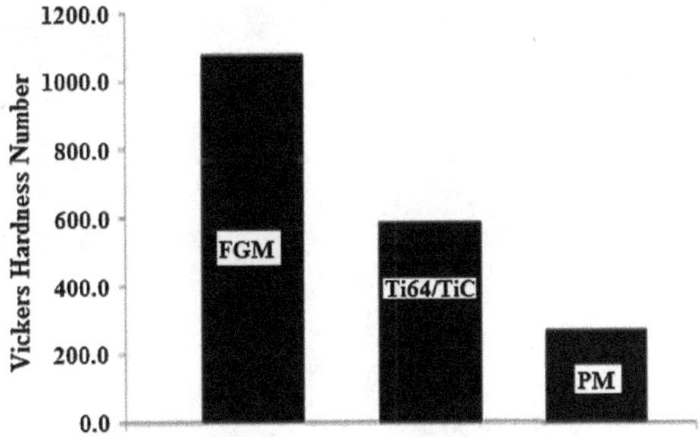

Fig. 5.9 The bar chart of the three samples

Table 5.5 Wear results

Sample designation	Wear volume (mm^3)
FGM	0.021
Ti64/TiC	0.030
PM	0.120

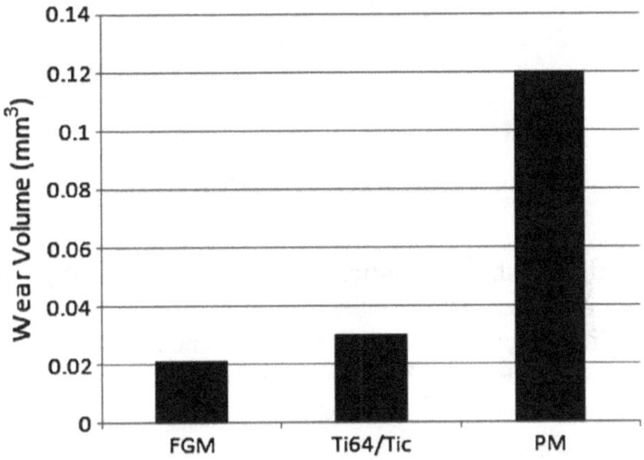

Fig. 5.10 The bar chart of wear volume for each of the samples

Fig. 5.10. The results showed that the FGM has the least wear volume. The micrograph of the conventional composite is shown in Fig. 5.11.

Comparing the microstructure of the FGM sample shown in Fig. 5.6 and that of the Ti64/TiC composite shown in Fig. 5.11; it can be seen that there is no sharp interface in the FGM; while the sharp interface in the Ti64/TiC is evident. Figure 5.12a Shows the microstructure of the FGM in the mid-region. The microstructure showed a combination of fully melted TiC powder and some unmelted TiC powder particles. The EDX results, as shown in Fig. 5.12b,

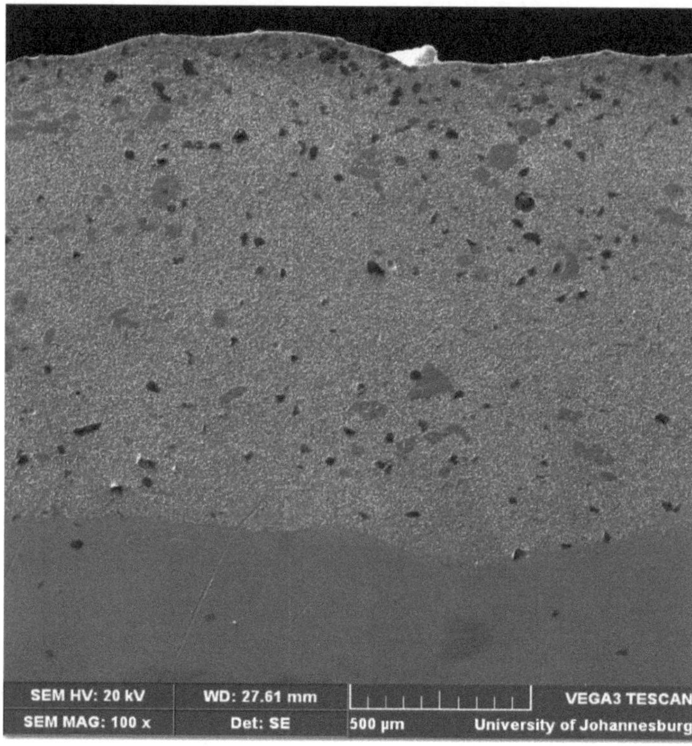

Fig. 5.11 The micrograph of the Ti64/TiC composite

confirmed that the particle shown in the labels A, B, and C in the Fig. 5.12a. These are unmelted TiC powder particles.

The microstructure of the fully melted part of the FGM in the mid-region is shown in Fig. 5.13a. The same microstructure of the sample in Fig. 5.13a at a higher magnification, is shown in Fig. 5.13b, in order to see the microstructure more clearly.

The fully melted TiC powder, shown in Fig. 5.13b, is dendritic in nature. This is as a result of the rapid solidification that is taking place during the LMD process. The microstructure of the lowest part of the FGM is shown in Fig. 5.14. It can be seen from these two microstructures that the size and the quantity of the UMC particles are greater in Fig. 5.14b than in Fig. 5.14a. The reason for this observation may be attributed to the fact that there is more TiC in Fig. 5.14b, which could be responsible for the presence of the large UMC TiC particles. The high magnification of the same micrographs in Fig. 5.14a is shown in Fig. 5.15a; and the higher magnification of the microstructure is shown in Fig. 5.15; and it is shown in Fig. 5.15b.

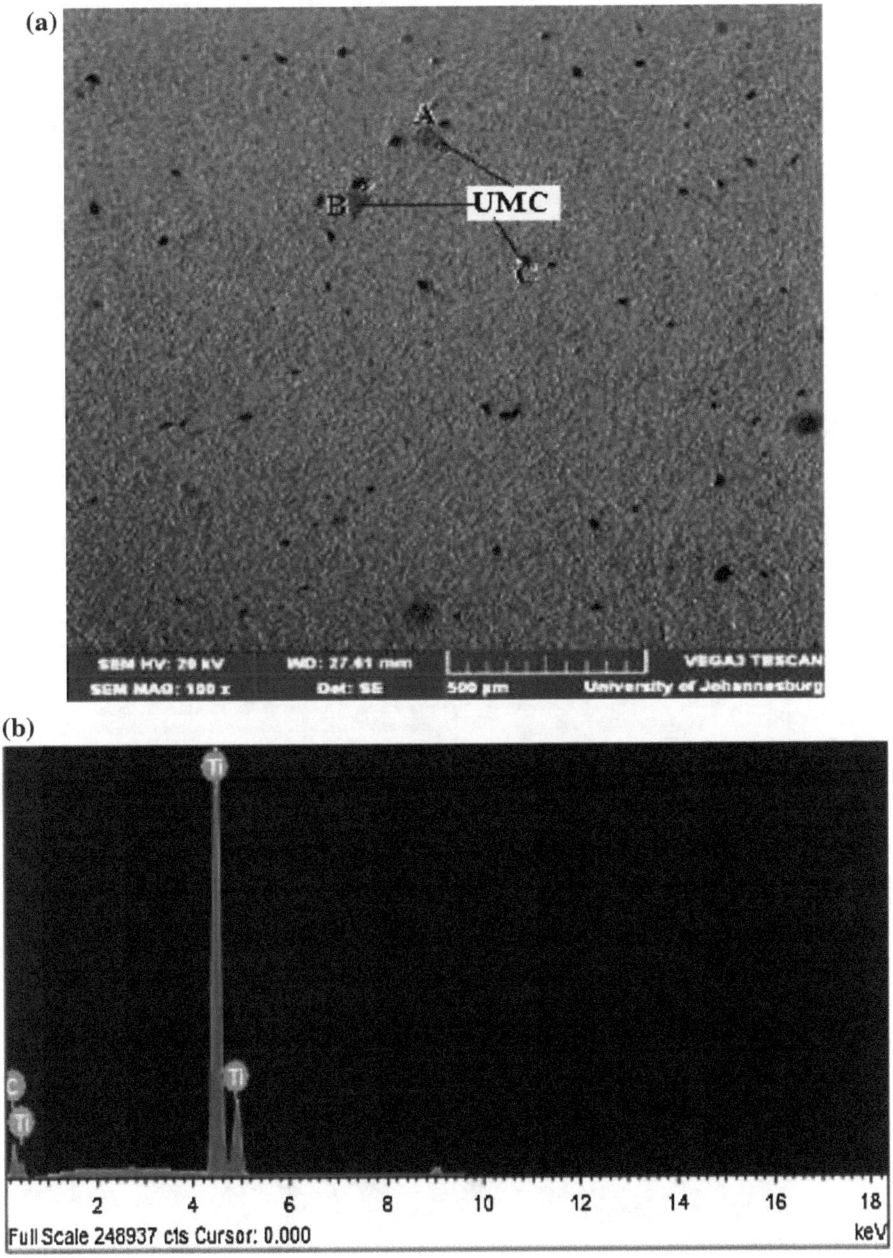

Fig. 5.12 The microstructure of the FGM in the mid-region shown in **a** and the EDX of the labels A, B, and C [18]

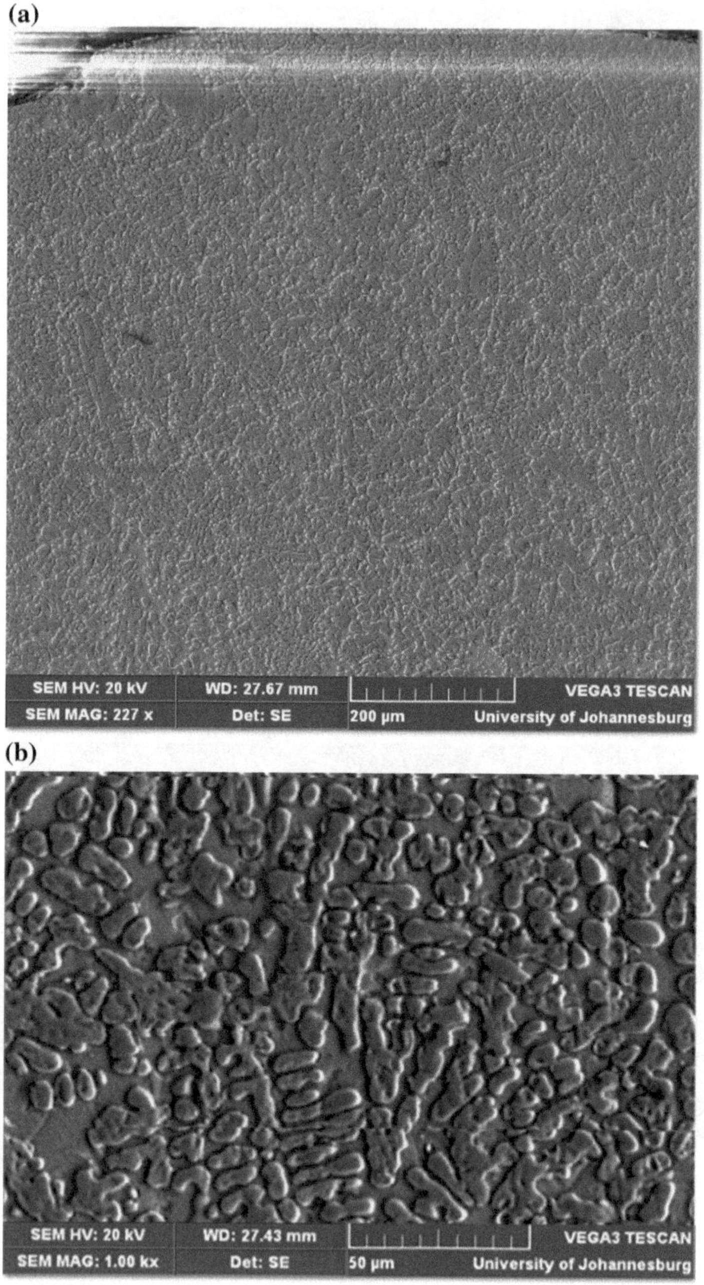

Fig. 5.13 The microstructure of the middle part of FGM sample

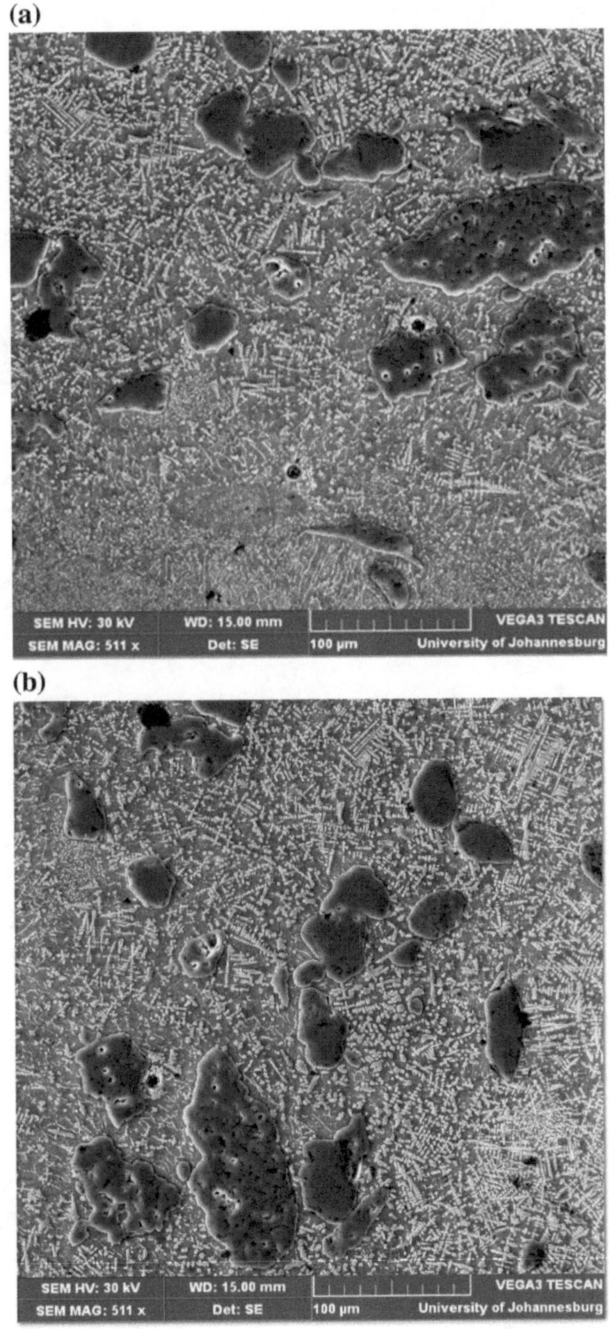

Fig. 5.14 The micrograph of the part labelled. **a** ‘b’ and **b** ‘c’ in Fig. 5.6

Fig. 5.15 The micrograph of sample in Fig. 5.14a at **a** high magnification; and **b** higher magnification

(a)

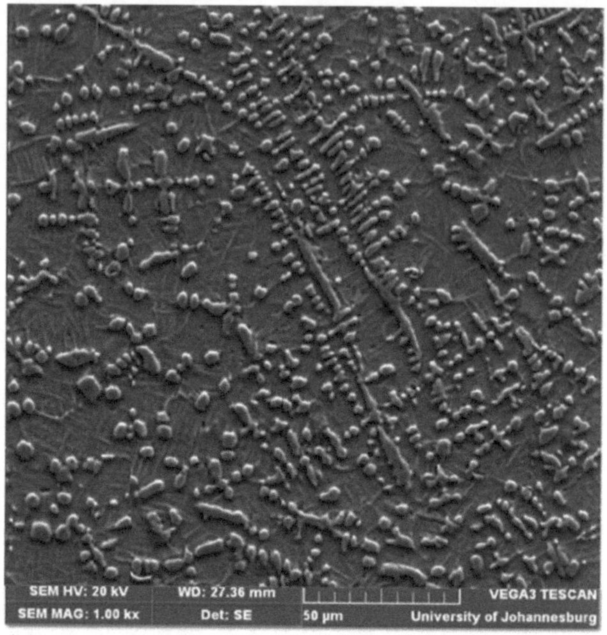

(b)

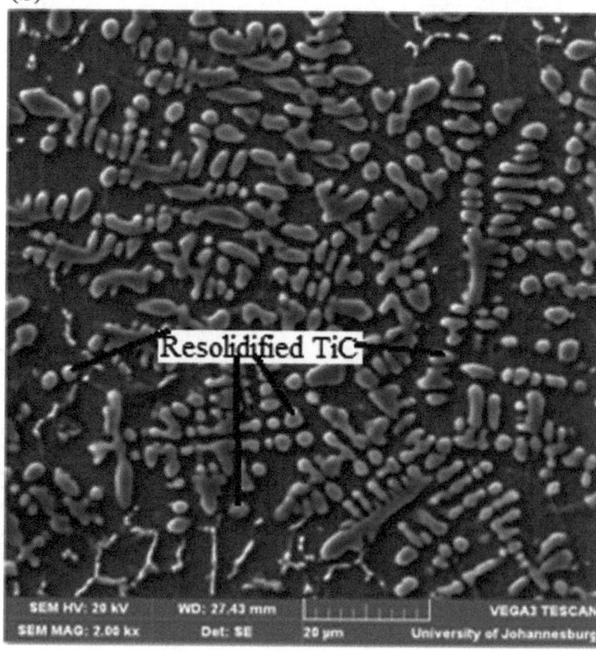

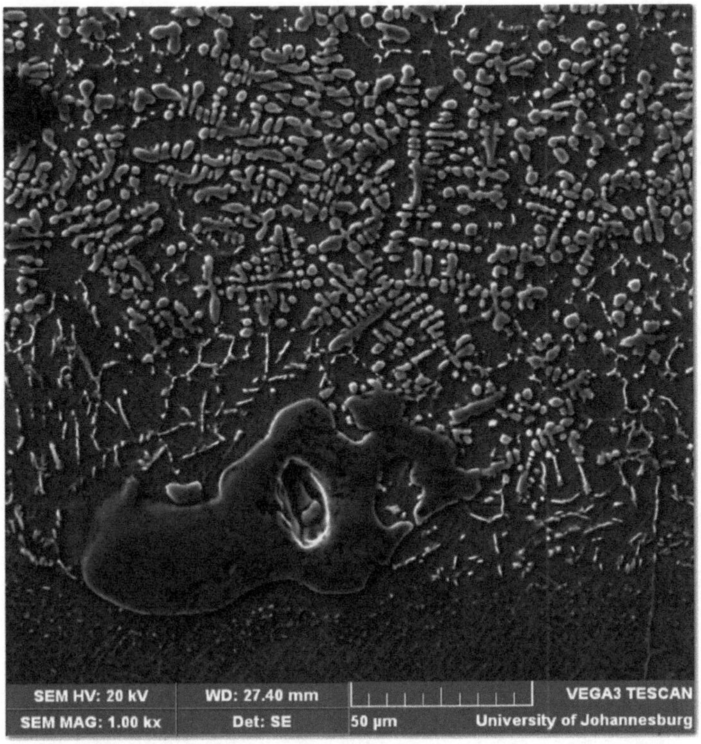

Fig. 5.16 The micrograph of the Ti64/TiC composite

The microstructure shown in Fig. 5.15 showed a combination of dendritic TiC and resolidified carbide-TiC particles.

This similar micrograph is also shown in Fig. 5.16, the micrograph of the lower part of the Ti64/TiC composite. The sharp interface is clearly seen, with the dendritic TiC and the resolidified carbides. The improvement in the wear-resistance behaviour of the FGM is seen through the study of the wear track. The wear tracks of the PM, the Ti64/TiC, and the FGM are shown in Figs. 5.17, 5.18, and 5.19, respectively. The substrate shows the highest wear rate, when compared with the Ti64/TiC composite and the FGM. The wear track of the substrate shown in Fig. 5.17a, b confirms this observation by looking at the wear mechanism in this Figure. The wear mechanism is seen as a combination of abrasion, adhesion, and plastic deformation, as seen in the wear scar. The abrasion wear mechanism began immediately the tungsten carbide ball was engaged in the rubbing action with the substrate.

The frictional force between the Titanium alloy and the tungsten carbide ball increases, as the rubbing action progresses—due to the chemical behaviour of the

(a)

(b)

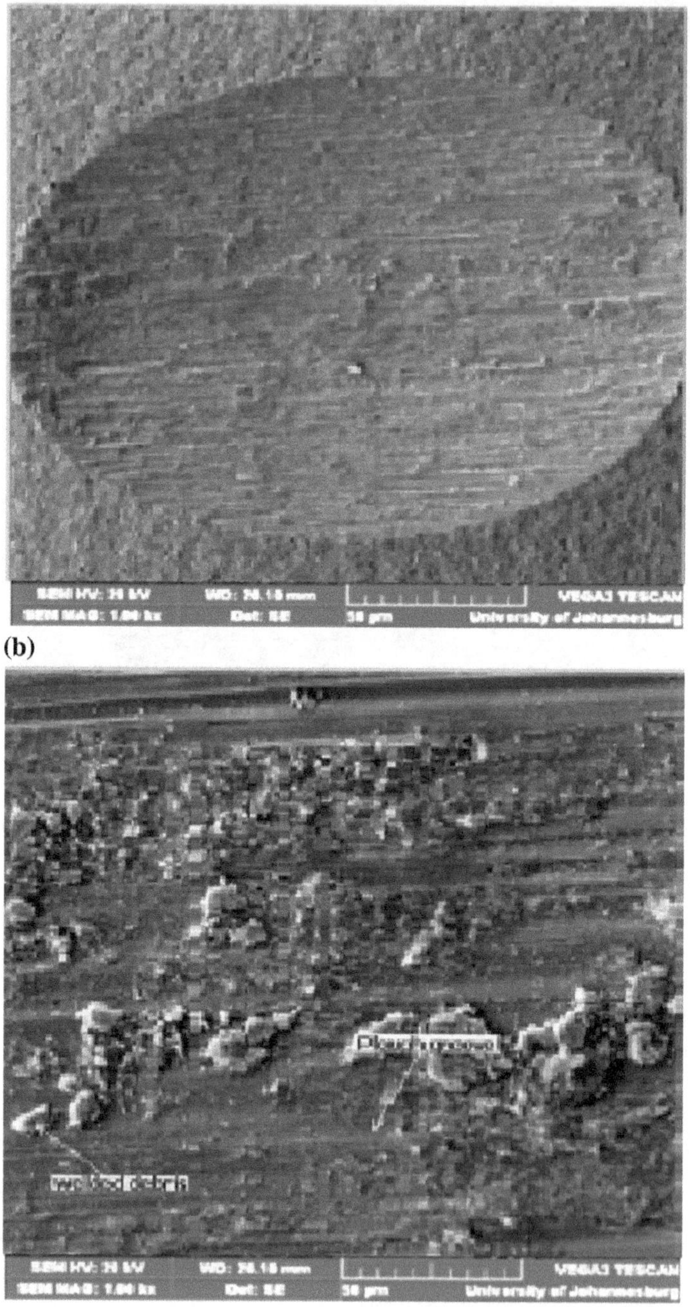

Fig. 5.17 The SEM micrograph of the wear track of the substrate at **a** low magnification and **b** higher magnification [28]

Fig. 5.18 SEM micrograph of the wear track of Ti64/TiC composite

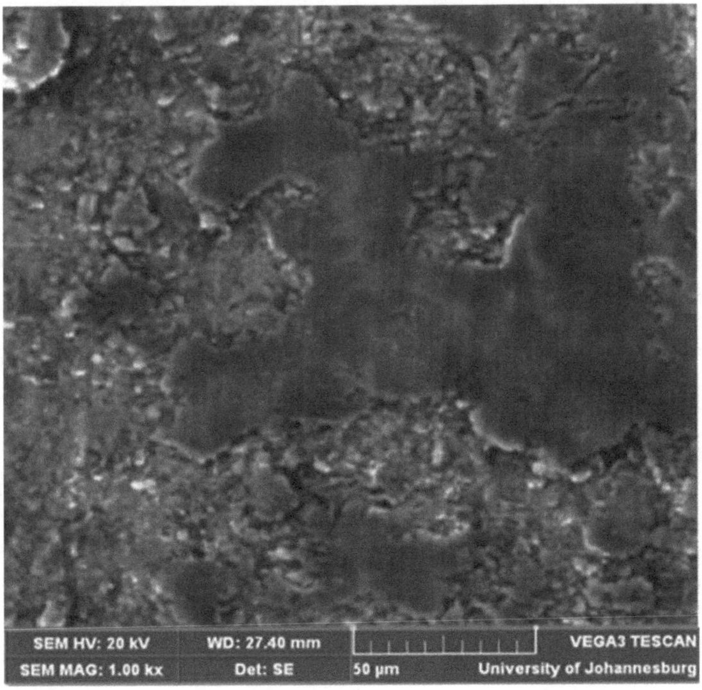

Fig. 5.19 SEM micrograph of the wear track of the FGM

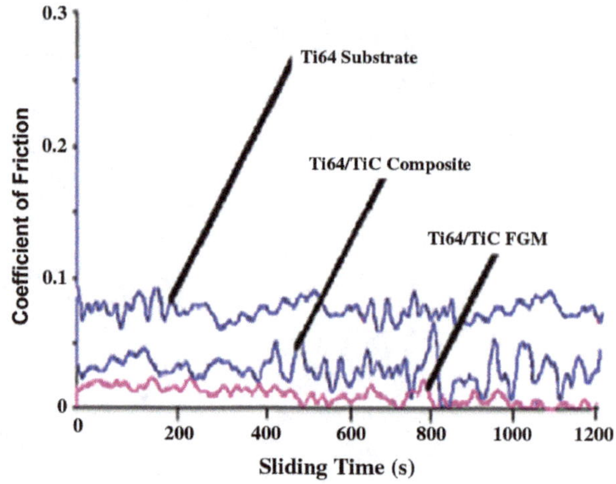

Fig. 5.20 The graph of the coefficient of friction against the sliding time

Ti64, and its high coefficient of friction (see Fig. 5.20). This high coefficient of friction causes the frictional force between the two rubbing materials to be high; and it causes a strong adhesion of these two surfaces. This high frictional force causes the temperature between the two surfaces to be increased. The high temperature is trapped in between the two surfaces, because of the poor thermal behaviour of titanium that later resulted in plastic deformation. As the rubbing action continues, the plastically deformed titanium alloy is chipped and forms debris in the process.

Some of the debris produced tends to adhere to the surface of the counter body, that is, the tungsten carbide ball. The remaining debris remains in between the sliding faces. The debris found in between the contact surfaces changed the wear mechanism from the two-body wear mechanism into a three-body wear mechanism [28]. The debris becomes work-hardened, as the wear action continues, which causes this debris to aggravate the wear action by cutting and tearing the surface of the substrate. These cutting and tearing results in the characteristic parallel grooves are to be seen on the wear track, as shown in Fig. 5.17b. The wear track of the Ti64/TiC composite is shown in Fig. 5.18; and this showed a better wear-resistant property. The improvement in the wear-resistant property, as shown by the wear scar, can be attributed to the presence of TiC in the composite.

The better wear-resistant property observed in the FGM, when compared with the Ti64/TiC composite is as a result of the varying amounts of UMC particles at different levels of the FGM. The quantity of the UMC particles in the Ti64/TiC composite remains constant. The high microhardness value of the FGM is also responsible for the improved wear-resistant behaviour of the FGM sample. Although, the wear volume produced by the composite is better than that of the substrate; the UMC particles first caused some cutting, before being ground into a

Fig. 5.21 XRD analysis showing the phases present in the FGM sample [18]

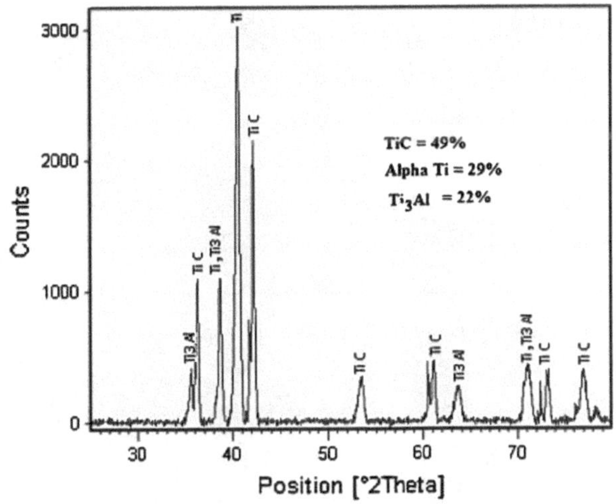

fine powder. That was why the wear scar contained some deeper grooves than the FGM wear scar did.

The addition of TiC to Ti64 helps to reduce the coefficient of friction; as can be seen in Fig. 5.20. The wear scar of the FGM sample showed the best wear-resistant property; as can be seen in Fig. 5.19. The wear track showed that the UMC particles play a big role in improving the wear-resistant behaviour. The UMC particles chipped off during the sliding wear action; and they are rubbed against other UMC particles, as well as the two contact surfaces. This results in the grinding of the UMC particles, which reduced their sizes into very fine powder particles. The fine powder particles helped to form a powder lubricant between the two surfaces; as the sliding progresses; and hence, inhibiting the wear action.

The XRD analysis showed the different phases present in the FGM sample, as shown in Fig. 5.21. The intermetallic titanium aluminide-Ti_3Al could be observed in these phases. The intermetallic compounds are usually very hard and could be detrimental in some applications. The intermetallic compounds found in this FGM were seen to help in the improvement of the wear-resistant property of the FGM sample.

The SEM micrograph of the FGM sample showing the intermetallic compound and the EDX analysis as shown in Fig. 5.22a, b, respectively.

This serves as a reinforcement; and it helped to improve the wear-resistant property of the FGM composite. The intermetallic compound of the Ti_3Al has been proved in the literature to have helped to improve the wear-resistant property of some metals [29, 30].

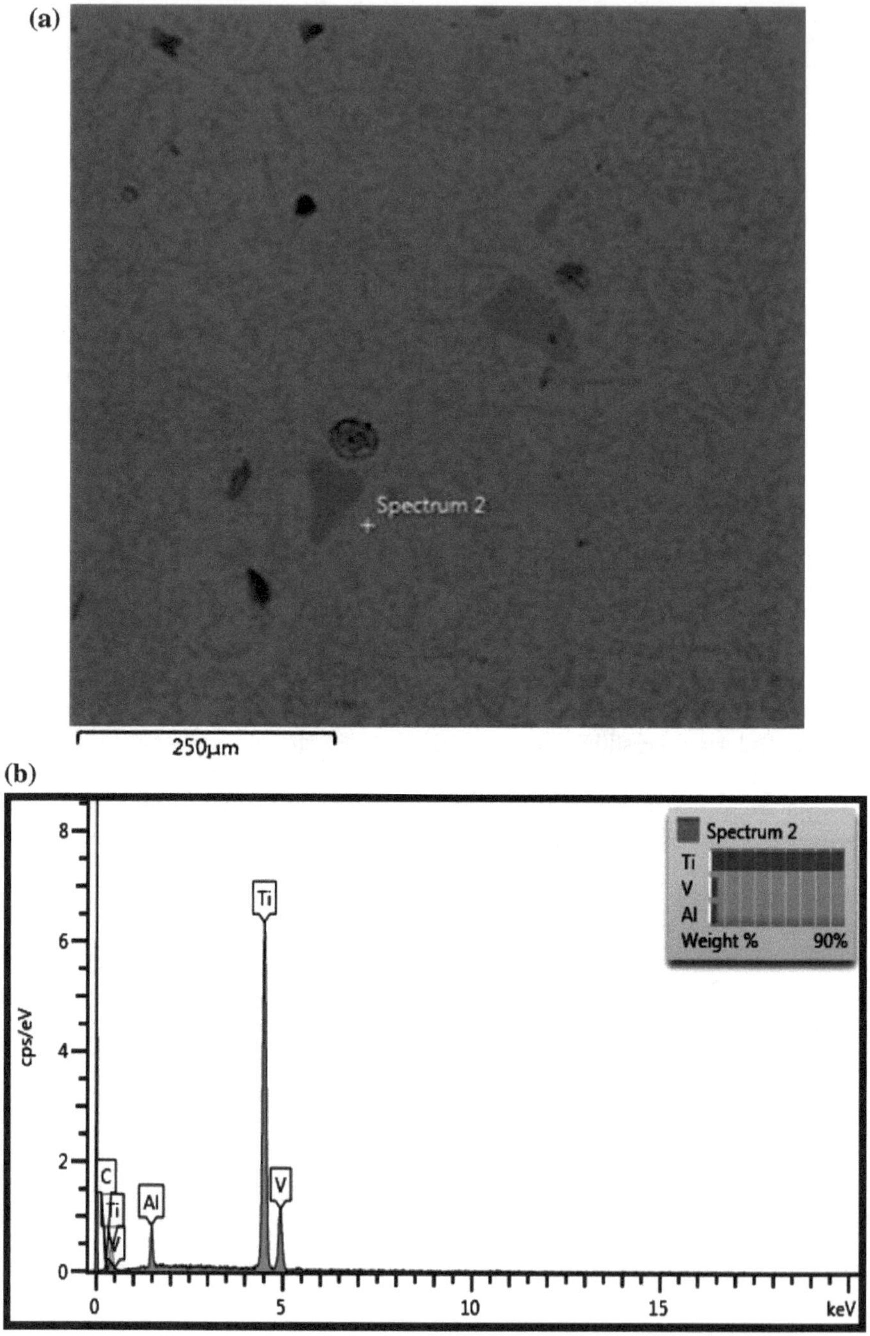

Fig. 5.22 **a** The SEM microstructure of the optimized FGM sample showing the intermetallic compound. **b** EDS analysis of the spectrum labelled 2 in **a** [18]

5.4 Summary

A case study of the fabrication of FGM, by using the LMD process has been presented in this chapter. The FGM produced consists of eleven layers of varying content of TiC, from 0% TiC: 100% Ti64, 5% TiC: 95% Ti64 up to 50% TiC: 50% Ti64. In order to compare the properties of the FGM produced, a conventional composite of Ti64/TiC composite was also produced, using the LMD process. A 50% TiC: 50% Ti64 of eleven layers was produced. The microstructure, microhardness, and wear-resistant properties of the FGM, the Ti64/TiC composite and the Ti64 substrate were investigated and presented in this chapter. The results of the properties of the three samples were compared; and it was discovered that the FGM composite of Ti64/TiC has the best microhardness and wear-resistant properties, when compared with the traditional Tu64/TiC composite and the Ti64 substrate. The microstructural analysis of the three samples showed that the conventional composite of Ti64/TiC has a distinct interface between the deposit and the substrate; while the FGM does not show any distinct interface; but rather a continuous microstructure, with a gradually increasing TiC content.

It may be concluded that the functionally material produced has better properties; and that the laser metal deposition process is a suitable manufacturing process for continuous functionally graded materials.

Acknowledgements This work is supported by University of Johannesburg Research Committee Fund, , the Department of Higher Education and Training (DHET) South Africa, the National Laser Centre Rental Pool Programme (RPP) contract number NLC - LREHA02-CON-001 and L'Oreal-UNESCO For Women in Science.

References

1. Cui, Z.D., Zhu, S.L., Man, H.C., Yang, X.J.: Microstructure and wear performance of gradient Ti/TiN metal matrix composite coating synthesized by using a gas-nitriding technology. Surf. Coat. Technol. **190**(2–3), 309–313 (2005)
2. Ramesh, S., Karunamoorthy, L., Palanikumar, K.: Surface roughness analysis in machining of titanium alloy. Mater. Manuf. Process. **23**(2), 174–181 (2008)
3. Peters, M., Kumpfert, J., Ward, C. H., Leyens, C.: Titanium alloys for aerospace applications. Adv. Eng. Mater. **5**(6), 419–427 (2003)
4. Straffelini, G., Andriani, A., Tesi, B., Molinari, A., Galvanetto, E.: Lubricated rolling-sliding behaviour of ion nitrides and untreated Ti.6Al.4V. Wear **256**(3–4), 346–352 (2004)
5. Zhang, S., Wu, W.T., Wang, M.C., Man, H.C.: In situ synthesis and wear performance of TiC particle reinforced composite coating on alloy Ti6Al4V. Surf. Coat. Technol. **138**(1), 95–100 (2001)
6. Arrazola, P.J., Garay, A., Iriarte, L.M., Armendia, M., Marya, S., Le Maître, F.: Machinability of titanium alloys (Ti6Al4V and Ti555.3). J. Mater. Process. Technol. **209**(5), 2223–2230 (2009)
7. Scott, J., Gupta, N., Wember, C., Newsom, S., Wohlers, T., Caffrey, T.: Additive manufacturing: status and opportunities, Science and Technology Policy Institute. Available from: https://www.ida.org/stpi/occasionalpapers/papers/AM3D_33012_Final.pdf (2012). Accessed 11 July 2016

8. Obielodan, J., Stucker, B.: Characterization of LENS-fabricated Ti6Al4V and Ti6Al4V/TiC dual-material transition joints. Int. J. Adv. Manuf. Technol. **66**(9–12), 2053–2061 (2013)
9. Popoola, A.P.I., Ochonogor, O.F., Abdulwahab, M.: Corrosion and hardness characteristics of laser-surface modified Ti6Al4V/Zr + TiC and Ti6Al4V/Ti + TiC Composites. Int. J. Electrochem. Sci. **8**, 2449–2458 (2013)
10. Wang, F., Mei, J., Jiang, H., Wu, X.: Laser fabrication of Ti6Al4V/TiC composites, using simultaneous powder and wire feed. Mater. Sci. Eng., A **445–446**, 461–466 (2007)
11. Ochonogor, O.F., Meacock, C., Abdulwahab, M., Pityana, S., Popoola, A.P.I.: Effects of Ti and TiC ceramic powder on laser-cladded Ti–6Al–4V in situ intermetallic composite. Appl. Surf. Sci. **263**, 591–596 (2012)
12. Foroozmehr, E., Sarrafi, R., Hamid, S., Kovacevic, R.: Synthesizing of functionally graded surface composites by laser powder deposition process for slurry erosion applications. Proceedings of the 20th Solid Freeform Fabrication Symposium, Austin (2009)
13. Balla, V.K., De Vas ConCellos, P.D., Xue, W., Bose, S., Bandyopadhyay, A.: Fabrication of compositionally and structurally graded Ti–TiO2 structures using laser-engineered net shaping (LENS). Acta Biomater. **5**(5), 1831–1837 (2009)
14. Thivillon, L., Bertrand, P.H., Laget, B., Smurov, I.: Potential of direct metal-deposition technology for manufacturing thick functionally graded coatings and parts for reactor components. J. Nucl. Mater. **385**(2), 236–241 (2009)
15. Zhang, Y., Wei, Z., Shi, L., Xi, M.: Characterization of laser powder deposited Ti–TiC composites and functionally graded materials. J. Mater. Proc. Technol. **206**, 438–444 (2008)
16. Wang, F., Mei, J., Wu, X.: Compositionally graded Ti6Al4V + TiC made by direct laser fabrication, using powder and wire. Mater. Des. **28**, 2040–2046 (2007)
17. Liu, W., DuPont, J.N.: Fabrication of functionally graded TiC/Ti composite by laser-engineered net shaping. Scripta Mater. **48**(9), 1337–1342 (2003)
18. Mahamood, R.M., Akinlabi, E.T.: Laser-metal deposition of functionally graded Ti6Al4V/TiC. Mater. Des. **84**, 402–410 (2015)
19. Shah, K.: Laser direct-metal deposition of dissimilar and functionally graded alloys. PhD Thesis, The University of Manchester (2011)
20. Lin, X., Yue, T.M., Yang, H.O., Huang, W.D.: Laser-rapid forming of SS316L/Rene88DT graded material. Mater. Sci. Eng., A **391**(1–2), 325–336 (2005)
21. Qin, L.Y., Yang, G., Wang, W., Tong, M., Wang, X.L.: Application of laser-metal deposition for the fabrication of titanium matrix wear-resistant coating and its wearing performance. Appl. Mech. Mater. **44–47**, 316–320 (2010)
22. Schwendner, K.I., Banerjee, R., Collins, P.C., Brice, C.A., Fraser, H.L.: Direct laser deposition of alloys from elemental powder blends. Scripta Mater. **45**(10), 1123–1129 (2001)
23. Mahamood, R.M., Akinlabi, E.T., Shukla, M., Pityana, S.: Characterization of laser-deposited Ti6A4V/TiC Composite. Lasers Eng. **29**(3–4), 197–213 (2014)
24. Standard ASTM.: E3, Standard Guide for Preparation of Metallographic Specimens, vol. 3, no. 1. ASTM International Book of Standards (2011)
25. ASTM E384-11e1: Standard Test Method for Knoop and Vickers Hardness of Materials, vol. 3, no. 1. ASTM International Book of Standards (2011)
26. ASTM G133-05: Standard Test Methods for Linearly Reciprocating Ball-on-Flat Sliding Wear, vol. 3, no. 2. ASTM International Book of Standards (2010)
27. Sharma, S., Sangal, S., Mondal, K.: On the optical microscopic method for the determination of ball-on-flat surface linearly reciprocating sliding-wear volume. Wear **300**(1–2), 82–89 (2013)
28. Mahamood, R.M., Akinlabi, E.T., Shukla, M., Pityana, S.: Scanning velocity influence on microstructure, microhardness and wear-resistance performance on laser-deposited Ti6Al4V/TiC composite. Mater. Des. **50**, 656–666 (2013)
29. Shimaa, E., Hisashi, S., Eri, M., Yoshimi, W.: Fabrication of Al-Al_3Ti/Ti_3Al functionally graded materials under a centrifugal force. Materials **3**, 4639–4656 (2010)
30. Li, J., Chen, C., Zhang, C.: Phase constituents and microstructure of Ti_3Al/Fe_3Al + TiN/TiB_2 composite coating on titanium alloy. Surf. Rev. Lett. **18**, 103–108 (2011)

Chapter 6
Future Research Direction in Functionally Graded Materials and Summary

6.1 Introduction

Functionally graded materials (FGMs) demonstrate a gradual variation in the material and/or the microstructural composition in relation to the spatial position in the material. This results in a variation in the properties that is suitable for the series of applications, such as the aerospace, the biomaterials, the defence, the electronics, the power engineering, and the other engineering applications [1–5]. The functionally gradedmaterial (FGM) is a position-dependent tailoring of materials that are designed, according to the functionality requirement of the material. The FGM has evolved from the composite materials that have been widely employed as a solution in the various engineering problems in the past.

The functionally gradedmaterial (FGM) was developed to solve the limitations that were posed by the conventional composite material. This provided an excellent opportunity to have the best of two, or more, material properties in a way that was not possible in the past. The FGM could be applied as a thin coating, or as a bulk material, depending on the intended application area. Apart from the gradation in material composition and graded microstructure, FGMs could also be developed through the gradation in porosity in the material [6–10].

The gradient in the structure of the material results in an overall change in the material properties, such as the elastic modulus, the thermal expansion behaviour, the reduced density and the improved hardness of materials. The design of FGM helps in the creation of materials with superior and multiple properties—without any form of weak interface [11].

The actual concept of FGM was acquired from nature and used to solve engineering problems in the same way that nature has used such materials based on their application requirement and areas of application. The application of FGM in thin film coatings has helped to reduce stress, prevent the peeling of the coated layer with time; and it also helps to prevent microcrack formation and the proliferation of cracks. The traditional manufacturing process of bulk FGM is usually divided into

R.M. Mahamood and E. Titilayo Akinlabi, *Functionally Graded Materials*,
Topics in Mining, Metallurgy and Materials Engineering,
DOI 10.1007/978-3-319-53756-6_6

two main steps, namely: the preparation of the gradient structure—that is, the building up of the spatially heterogeneous structure by arranging the various compositions, according to the FGM design; and the consolidation process, which involves the transformation of this structure into a dense bulk FGM structure, which could be achieved through any densification process, such as sintering.

The FGM can also be produced through the use of an advanced manufacturing process known as the additive manufacturing process. The main advantage of the additive manufacturing process is that it can be used to produce FGM—no matter the complexity—in a single manufacturing run [12, 13]. The background, types, manufacturing methods, the use of additive manufacturing technology for producing FGM, and a case study on the production of FGM using the laser metal deposition process were presented in Chaps. 1–5, respectively.

In this chapter, the future research direction in the field of FGM is presented. The chapter ends with a comprehensive summary of the book as a whole.

6.2 Future Research Need in Functionally Graded Materials

A lot of research work has been undertaken since the inception of the FGM; and some of this includes the various work done by distinguished researchers [13–35]. Much of such research work was dedicated to the design of FGM [20–24]. Some research work was conducted based on the modelling of the FGM [18, 25–28]. The fabrication of the FGM has been the focus of some of the research work in the literature [29–34]. The characterization of FGMs has been extensively studied by a number of researchers [12, 35–45]. While much of the review work has also been done on various parts of FGM; some of the notable work reported in these research projects can be read [13, 16, 18, 46–51].

Considerable progress has been made since the discovery of FGM; and the works are still ongoing, in order to resolve some of the issues with these important engineering materials. Suggestions for future research needed by some of these researchers are summarized in this section [13–19].

Mahamood et al. [13] presented an overview of FGM. These FGMs were seen as excellent advanced engineering materials with a great potential to revolutionize the manufacturing world in the twenty-first century. Some of the manufacturing processes for the FGM were reviewed in this work; and it was concluded that a number of setbacks exist in terms of the fabrication cost of the FGM since this is still very high at the moment. The high cost of FGM was blamed on the high cost of the powder processing and on the fabrication methods. The promising fabrication method that is capable of reducing some of the financial burden in the production of FGM is the Additive manufacturing (AMP) process. Additive manufacturing technologies offer greater advantages for producing the FGM; because the technology can be used to fabricate any part—no matter how complex—and with the required gradation in properties, by simply adding materials layer after layer by following the data from the three-dimensional (3D) computer-aided design (CAD) data information.

In spite of the great potentials of this revolutionary technology [52], there are still a lot of unresolved issues that need to be resolved with this great technology. It was suggested that more research work needs to be performed on improving the performance of these AM processes through extensive material characterization of the functionally grade materials. This is to enable the generation of the needed comprehensive database for the process; and also to develop a proper predictive model that would help in the control of the AM process.

It has also been suggested that further research work needs to be conducted to improve the AM process control through the development of a more powerful feedback control algorithm for the overall FMG fabrication process. It is believed that by achieving all these proposed research aims; it would help to bring down the cost of production of the FGMs, as well as to improve the reliability in the material, as well as in the fabrication process.

Saiyathibrahim et al. [14] reviewed the various processing techniques of the FGM. An experimental investigation was used for these processing techniques using different materials. Some of the application areas of the FGM are also presented in this study. The role played by some of the processing parameters and their influence on the resulting properties of the FGM produced are also reviewed in this work.

Some of the properties that were analysed include: the microstructural evaluation, the wear mechanisms, the porosity, and the stress distributions in FGM materials, such as the metal–metal, the metal–ceramic and the ceramic–ceramic interfaces. The study revealed that the performance of processing techniques needs to be improved; and there is an increased need for extensive research on the material characterization of FGM components. These were suggested with the hope of bringing down the cost of manufacturing the FGMs; and also help to increase the productivity. The future research need, as mentioned by the authors, includes the consideration of the industrial needs of new FGM families. Additionally, their processing methods need to be further researched—especially with regard to the real applications of the functionally graded material. It was also pointed out that the FGM gradation preparation is the critical step in the FGM production; and the process needs to be improved by using computer-assisted modelling, which would help to forecast the properties of the gradient formation and hence help to reduce the overall production cost. With this development, the calculations related to the properties and the desired combinations of the FGM properties would have been done before the actual FGM production.

Material characterization by way of non-destructive methods are also suggested, as a future research need for functionally graded material. Upscaling of laboratory studies into an industrial scale is also proposed. The work concluded that there is still a lot of research needed to adapt the existing processing routes to an industrial scale. Proper technology transfer of the FGM from the laboratory to the industry is proposed as the key to the needed development and the needed cost reduction in the fabrication of the functionally graded material.

The reproducibility in geometry, gradation and the properties of FGMs are also some of the other research needs suggested by the authors. In a similar study by the authors, the consideration of industrial needs of new FGM families and their effective processing methods have also been proposed [15]. Udupa et al. [16] proposed the need for future research work in the area of changing the hardness of FGM composite because of the potential future role this will play in the future of aerospace and space applications. It was stated that FGM requires multidisciplinary, multi-industrial and international collaboration to achieve the desired milestone in the FGM industry.

A case study was presented in this research work on the influence of carbon nanotube reinforcement in FGM, in order to change the physical properties of the FGM produced. The need for further research work on material characterization has also been stressed by Lee [17]. It has also been proposed that there is research needed for improving the efficiency of the solar roofing panel system, a functionally graded material. Additionally, the need of a novel, economical and sustainable manufacturing method for FGM has also been suggested for high-quality material properties.

It was concluded that the application of FGM could be extended to some of the promising materials of structures and infrastructures for the functional purposes of sound resistance, heat insulation and fireproofing applications; if the production cost of the FGM can be reduced through further research work.

Birman and Byrd [18] reviewed the research progress in the characterization, modelling, and analysis of FGM. The study revealed that more is still needed to be done in terms of research, in order to address a number of issues on the FGM. Some of their suggestions for the future research needed are presented as follows:

- Homogenization of particulate FGM, where the material has a significant gradient since the interaction between the particles should not be disregarded. The material properties that are often evaluated, according to the theoretical models often disagree with the measured values of FGM constants. This shows that there is a need for a probabilistic approach to homogenization that would help to account for uncertainty in the actual material distribution throughout the volume.
- It was also suggested that in the event of a probabilistic homogenization problem, there is the need for the application of probabilistic mechanics for the analysis of the response, fracture, and fatigue characteristics of the FGM structures.
- Although there are wide range of FGM manufacturing techniques, more is still needed in terms of research on the procedures and protocols that would help to guarantee a reliable and predictable distribution of the material constituent phases and properties throughout the structure.
- There is a need for research that would help to account for the influence of temperature on the material properties. This, in turn, would affect the solution of the heat conduction. The problems encountered in the formulation of FGM, accounting for the temperature-dependent FGM properties could be solved by

using any iterative technique. The exact solution may also be found in a relatively simple benchmark case.

- There is research needed that would help to account for the thermal residual stresses in the FGM analysis since the residual stresses have a great influence on the strength and the fracture resistance of these FGM materials. In the fabrication process that does not involve layered construction, the residual stresses—in terms of microscopic stresses resulting from a thermal mismatch of the constituent materials—should be of great concern, and this needs to be accounted for. The macromechanical residual stresses may also occur if the post-processing deformations are restrained in the functionally graded material. That is, if the properties change in a stepwise manner. The macromechanical residual stresses due to the different layers and their different thermal expansion coefficients should be incorporated in the analysis.
- The influence of temperature on the stress, stability and vibration problems of the FGM should also be accounted for, as well as any changing in the material properties, as a result of temperature changes. For example, if one of the constituent material phases is more sensitive to temperature changes than the other phases, the degree of property changes would not be uniform throughout the material, even though it is subjected to a uniform temperature.
- There is further research needed in the area of optimization problems for the FGM because of the combination of dissimilar materials that are involved.
- The research on the analysis of compressed FGM structures should always account for prebuckling deformations; because, the response of the FGMs in the static stress and their dynamic behaviour are affected by the inherited asymmetry in the FGM. This implies that there is a coupling effect even in the quasi-isotropic FGM and that the classical Euler bifurcation buckling is said not to exist in thermally loaded FGM structures, except those with fully clamped boundaries.
- The need to also balance the merit of grading in the FGM and the demerit of achieving such gradation is also stressed by the authors. This is because, the asymmetry that is being introduced during the gradation of the FGM often results in higher deformations and stresses in the FGM structure than in a homogeneous symmetric counterpart with an overall identical material composition. It is, therefore, important to emphasize the advantages and disadvantages that come with the design of FGM; and these should be accounted for during the material grading process.
- It was concluded that FGM is a promising engineering material for a diverse range of possible applications. However, a lot of research needs to be done for it to meet the necessary expectations. The research needs are seen to be quite numerous and diverse; but the FGM significant potential benefits would fully justify the necessary research efforts needed.

Shiota and Miyamoto [19] demonstrated that there is enormous research needed for the FGM to be able to live up to its expectation. The reason that was attributed to this is that the design functions are too difficult and very expensive to produce on an industrial scale. Furthermore, the available test methods are not sufficient to ascertain the quality, the reproducibility, the reliability and the life of the FGM. The following research needs were proposed: the basic research on the physical, chemical and mechanical behaviour of the FGMs and their relationship to the microstructure and their formation mechanism would go a long way in the ability to control the fabrication process. This is important, in order to tailor properly the functionally graded material. The urgent need of coherent scientific infrastructure in terms of the chemical, electronic and mechanical aspects of FGM has also been stressed. It was concluded that there are a number of research projects needed for the FGM, as follows: There is a need to define clearly the applications in need of FGMs, and the list of the required satisfactory properties for their functionality, identifying the candidate materials and evaluating the proper manufacturing route to be used to fabricate the needed FGM in terms of cost and effectiveness.

The quality criteria should be clearly stated, so that the performance of such FGMs developed can be measured against the laid down performance criteria. A non-disruptive method should also be developed to evaluate the properties of the FGM. Additionally, the environmental impact assessment on the recycling of the FGM should also be studied.

The major problem with the FGM currently is the manufacturing cost of FGM, and this has limited the application of these important engineering materials to the niche application areas. There is considerable research needed to develop a novel, economical and sustainable fabrication process for the production of the FGMs with high quality and reliable properties. This will no doubt extend the application area of this promising engineering material for such applications, as in structural and sound resistance applications.

6.3 Summary

This book has dealt with the functionally graded composite materials, an important advanced engineering material in six different chapters. The topic was introduced in Chap. 1. The brief background information on the FGM was presented; and some of the FGMs existing in nature were also discussed. The lessons from nature were used to develop the FGMs for use in the current life. Different materials are now being combined to produce FGMs for a variety of functions.

The different types of functionally graded materials and their areas of application were discussed in Chap. 2. There are different types of functionally graded materials that are in use today, depending on the type of application. The three main different types of FGMs that were described in this chapter are: the chemical composition graded FGM, the microstructure-graded FGM, and the porosity-graded FGM. The FGMs are used to solve a number of engineering problems, as well as for biomedical implants in human tissue replacements. These FGMs help to eliminate the stress singularities that occur, as a result of property mismatches. Although the chemical composition-graded FGM was initially developed for the thermal barrier application, the porosity gradient FGMs are now widely used as medical implants in the biomedical application because they are designed to have properties that are close to the natural human organs, which are found in nature. The high interest in this important material in the research community has caused it to evolve into different types. Additionally, the areas of application have also been expanded substantially. All the three types of FGMs and their various application areas are presented and discussed in this chapter. The application area can be summarized by a chart shown in Fig. 6.1.

Chapter 3 presents the fabrication methods for thin functionally graded coating and bulk functionally graded material. The two main methods for the production of the thin coating of FGM are discussed in Chap. 3. They are the physical vapour deposition and the chemical vapour deposition processes that are the two main fabrication methods for the production of functionally graded coatings that are well established in the literature, and they are now being used at the commercial level. The powder metallurgy and the casting-based fabrication technologies for the production of bulk FGM are also discussed in detail in this chapter. These manufacturing processes have been reviewed with their merits, demerits, and areas of application.

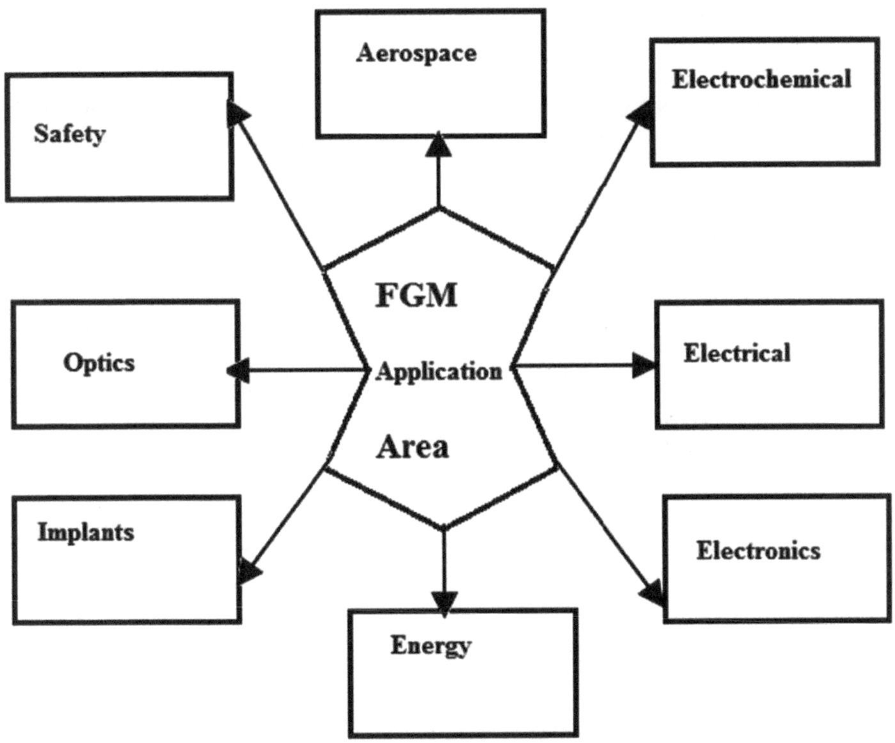

Fig. 6.1 Application of functionally graded materials

Chapter 4 presented the use of Additive manufacturing (AM) technology, a revolutionary manufacturing method, that produces 3D objects, regardless of their complexity directly from the three-dimensional (3D) computer-aided design (CAD) model of the object by adding materials layer by layer for the production of FGM. The different classes of AM technologies that can be used to fabricate FGM are discussed in this chapter. The different classes of AM presented are: The powder bed fusion and the material extrusion. The working principle of these additive manufacturing technologies was described; and some of the research efforts in this areas were reviewed. The main advantage of using AM technologies for the production of FGM is that FGM components can be produced, regardless of the complexity, and in a single manufacturing run. The researches in this area showed that improved properties in the FGM produced can be achieved by using these AM processes.

A case study of the fabrication of FGM using the LMD process was presented in Chap. 5. The case study of the FGM produced consists of eleven layers of varying content of TiC, from 0% TiC: 100% Ti64, 5% TiC: 95% Ti64 up to 50% TiC: 50% Ti64. To be able to compare the properties of the FGMs produced, a conventional composite of Ti64/TiC composite was also produced by using the LMD process consisting of A 50% TiC: 50% Ti64 and made of eleven layers. The properties that were studied and compared are the microstructure, the microhardness and the wear-resistant properties of the FGM, the Ti64/TiC composite and the Ti64 substrate.

The results showed that the FGM has the best microhardness and wear-resistant properties when compared with the Ti64/TiC composite and the parent material. The microstructural analysis results also showed that the conventional composite of Ti64/TiC has a distinct interface between the deposit and the substrate; while the FGM does not show any distinct interface, but a continuous microstructure with gradually increasing TiC content. It was concluded that the FGM has the best properties; and that the laser metal deposition (LMD) process is a suitable manufacturing process of continuous FGM materials.

In this chapter, which is Chap. 6, the future research direction in the field of FGM is presented. Suggested future works by various researchers are also presented. The chapter ends with the summary of the entire book.

Acknowledgements This work is supported by the University of Johannesburg Research Committee Fund, the Department of Higher Education and Training (DHET) South Africa, the National Laser Centre Rental Pool Programme (RPP) contract number NLC - LREHA02-CON-001 and L'Oreal-UNESCO For Women in Science.

References

1. Ekici, R., Apalak, M.K., Yildirim, M.: Indentation behaviour of functionally graded Al–SiC metal matrix composites with random-particle dispersion. Compos. B **42**, 1497–1507 (2011)
2. Siddharth, P. A., Bhatt, A.D.: Mechanical and dry sliding wear characterization of epoxy–TiO_2 particulates filled functionally graded composites materials using Taguchi design of experiment. Mater. Des **32**, 615–627 (2011)

3. Soon, Y.-M., Shin, K.-H., Koh, Y.-H., Lee, J.-H., Choi, W.-Y., Kim, H.-E.: Fabrication and compressive strength of porous hydroxyapatite scaffolds with a functionally graded core/shell structure. J. Eur. Ceram. Soc. **31**, 13–18 (2011)
4. Siddhartha, P.A., Bhatt, A.D.: Mechanical and dry sliding wear characterization of epoxy–TiO_2 particulate-filled functionally graded composite materials using the Taguchi design of experiment. Mater. Des. **32**, 615–627 (2011)
5. Ekici, R., Apalak, M.K., Yildirim, M.: Indentation behaviour of functionally graded Al–SiC metal matrix composites with random-particle dispersion. Compos. B **42**, 1497–1507 (2011)
6. Droschel, M., Hoffmann, M.J., Oberacker, R., Both, H.V., Schaller, W., Yang, Y.Y., Munz, D.: SiC-ceramics with tailored porosity gradients for combustion chambers. Key Eng. Mater. **175–176**, 149–162 (2000)
7. Cichocki Jr., F.R., Trumble, K.P., Rodel, J.: Tailored porosity gradients via colloidal infiltration of compression-moulded sponges. J. Am. Ceram. Soc. **81**, 1661–1664 (1998)
8. Harley, B.A., Hastings, A.Z., Yannas, I.V., Sannino, A.: Fabricating tubular scaffolds with a radial pore size gradient by a spinning technique. Biomaterials **27**, 866–874 (2006)
9. Bretcanu, O., Samaille, C., Boccaccini, A.R.: Simple methods to fabricate bioglass-derived glass ceramic scaffolds exhibiting porosity gradients. J. Mater. Sci. **43**, 4127–4134 (2008)
10. Li, R., Liu, J., Shi, Y., Du, M., Xie, Z.: 316L Stainless steel with gradient porosity fabricated by selective laser melting. J. Mater. Eng. Perform. (2009). doi:10.1007/s11665-009-9535-2
11. Kieback, B., Neubrand, A., Riedel, H.: Processing techniques for functionally graded materials. Mater. Sci. Eng. A **362**, 81–105 (2003)
12. Mahamood, R.M., Akinlabi, E.T.: Laser-metal deposition of functionally graded Ti6Al4V/TiC. Mater. Des. **84**, 402–410 (2015)
13. Mahamood, R.M., Akinlabi, E.T., Shukla, M., Pityana, S.: Functionally graded material: an overview. In: Proceedings of the World Congress on Engineering WCE 2012, vol. 3, pp. 1593–1597 (2012)
14. Saiyathibrahim, A, Mohamed, N.S.S., Dhanapal, P.: Processing techniques of functionally graded materials—a review. In: International Conference on Systems, Science, Control, Communication, Engineering and Technology, pp. 98–105 (2015)
15. Saiyathibrahim, A., Subramaniyan, R., Dhanapal, P. Centrifugally cast functionally graded materials—a review. In: International Conference on Systems, Science, Control, Communication, Engineering and Technology, pp. 68–73 (2016)
16. Udupa, G., Rao, S.S., Gangadharan, K.V.: Functionally graded composite materials: an overview. Procedia Mater. Sci. **5**, 1291–1299 (2014)
17. Lee, P-H.: Fabrication, characterization and modelling of functionally graded materials. Ph.D. Thesis, Columbia University (2013)
18. Birman, V., Byrd, L.W.: Modelling and analysis of functionally graded materials and structures. Appl. Mech. Rev. **60**, 195–216 (2007)
19. Shiota, I., Miyamoto, Y.: Functionally Graded Materials. Elsevier (1997)
20. Hedia, H.S., Aldousari, S.M., Abdellatif, A.K., Fouda, N.A.: New design of cemented stem using functionally graded materials (FGM). Biomed. Mater. Eng. **24**(3), 1575–1588 (2014). doi:10.3233/BME-140962
21. Zhang, Y., Sun, M.J., Zhang, D.: Designing functionally graded materials with superior load-bearing properties. Acta Biomater. **8**(3), 1101–1108 (2011). doi:10.1016/j.actbio.2011.11.033
22. Miyamoto, Y., Kaysser, W.A., Rabin, B.H., Kawasaki, A., Ford, R.G. (eds.): Functionally Graded Materials Design, Processing and Applications. Springer, US (1999)
23. Jing, S., Zhang, H., Zhou, J., et al.: Optimum weight design of functionally graded material gears. Chin. J. Mech. Eng. **28**, 1186–1193 (2015)
24. Kou, X.Y., Parks, G.T., Tan, S.T.: Optimal design of functionally graded materials, using a procedural model and particle swarm optimization. Comput. Aided Des. **44**(4), 300–310 (2012)
25. Markworth, A.J., Ramesh, K.S., Parks Jr., W.P.: Modelling studies applied to functionally graded materials. J. Mater. Sci. **30**(9), 2183–2193 (1995)

26. Birman, V.: Modelling and analysis of functionally graded materials and structures. In: Hetnarski, R.B. (ed.) Encyclopedia of Thermal Stresses, pp. 3104–3112. Springer, Netherlands (2014)
27. Gasik, M.M.: Micro-mechanical modelling of functionally graded materials. Comput. Mater. Sci. **13**(1–3), 42–55 (1998)
28. Rhee, R.S.: Multi-Scale modelling of functionally graded materials (FGMS) using finite element methods. Ph.D. Thesis, University of Southern California (2007)
29. Chumanov, I.V., Anikeev, A.N., Chumanov, V.I.: Fabrication of functionally graded materials by introducing Wolframium carbide dispersed particles during centrifugal casting and examination of FGM's structure. Procedia Eng. **129**, 816–820 (2015)
30. Liu, T., Wang, Q., Gao, A., Zhang, C., Wang, C., He, J.: Fabrication of functionally graded materials by a semi-solid forming process under magnetic field gradients. Scripta Mater. **57** (11), 992–995 (2007)
31. Zhu, K., Wang, H., Qiu, J., Luo, J., Ji, H.: Fabrication of 0.655Pb(Mg1/3Nb2/3)O_3-0.345PbTiO_3 functionally graded piezoelectric actuator by tape casting. J. Electroceram. **27**, 197–202 (2011)
32. Huang, X., Liu, C., Lv, X., Liu, G., Li, F.: Aluminum alloy pistons reinforced with SiC fabricated by centrifugal casting. J. Mater. Process. Technol. **211**, 1540–1546 (2011)
33. Kieback, B., Neubrand, A., Riedel, H.: Processing techniques for functionally graded materials. Mater. Sci. Eng., A **362**, 81–105 (2003)
34. Nemat-Alla, M.M., Ata, M.H., Bayoumi, M.R., Khair-Eldeen, W.: Powder-metallurgical fabrication and microstructural investigations of aluminum/steel functionally graded material. Mater. Sci. Appl. **2**, 1708–1718 (2011)
35. El-Wazery, M.S., El-Desouky, A.R., Hamed, O.A., Mansour, N.A., Hassan, A.A.: Preparation and mechanical properties of zirconia/nickel functionally graded materials. Arab J. Nucl. Sci. Appl. **45**(2), 435–446 (2012)
36. Jin, X., Wu, L., Sun, Y., Guo, L.: Microstructure and mechanical properties of ZrO_2/NiCr functionally graded materials. Mater. Sci. Eng. A. **509**, 63–68
37. Kiran, A.S., Narendranath, V.D., Mukunda, P.G.: Evolution of microstructure and hardness of Al–Si functionally graded material cast through centrifuge technique using hypereutectic and eutectic Al–Si. Int. J. Mech. Mater Eng (IJMME) **6**(2), 275–279 (2011)
38. Thirtha, P.H.P., Chikkanna, N.: Experimental investigation on the effect of particle loading on microstructural, mechanical and fractural properties of Al/Al_2O_3 functionally graded materials. Int. J. Adv. Eng. Technol. **2**(5), 161–166 (2011)
39. Vieira, A.C., Sequeira, P.D., Gomes, J.R., Rocha, L.A.: Dry-sliding wear of Al alloy/SiCp functionally graded composites: influence of processing conditions. Wear **267**, 585–592 (2009)
40. Siddhartha, A.P., Bhatt, A.D.: Mechanical and dry-sliding wear characterization of epoxy–TiO_2 particulate filled functionally graded composites materials using the Taguchi design of experiment. Mater. Des. **32**, 615–627 (2011)
41. Ekici, R., Apalak, M.K., Yildirim, M.: Indentation behaviour of functionally graded Al–SiC metal matrix composites with random-particle dispersion. Compos. B **42**, 1497–1507 (2011)
42. Zuiker, J.R.: Functionally graded materials: Choice of micro-mechanics model and limitations in property variation. Compos. Eng. **5**(7), 807–819 (1995)
43. Nakamura, T., Wang, T., Sampath, S.: Determination of properties of graded materials by inverse analysis and instrumented indentation. Acta Mater. **48**, 4293–4306 (2000)
44. Grujicic, M., Zhang, Y.: Determination of effective elastic properties of functionally graded materials using Voronoi cell finite element method. Mater. Sci. Eng. A **251**, 64–76 (1998)
45. Bohidar, S.K., Sharma, R., Mishra, P.R.: Functionally graded materials: a critical review. Int. J. Res (IJR). **1**(7) (2014)
46. Chauhan, P.K., Khan, I.A.: Review on analysis of functionally graded material beam type structure. Int. J. Adv. Mech. Eng. **4**(3), 299–306 (2014)
47. Kohli, G.S., Singh, T.: Review of functionally graded materials. J. Prod. Eng. **18**(2), 1–4 (2015)

48. Singh, L.: A review on functionally graded materials. Int. J. Res. Aeronaut. Mech. Eng. **4**(6), 8–14 (2016)
49. EL-Wazery, M.S., EL-Desouky, A.R.: Review on functionally graded ceramic-metal materials. Mater. Environ. Sci. **6**(5), 1369–1376 (2015)
50. Sharma, R., Jadon, V.K., Singh, B.J.: Inst. Eng. India Ser. C. **96**, 73. doi:10.1007/s40032-014-0125-1
51. Jha, D.K., Kant, T., Singh, R.K.: A critical review of recent research on functionally graded plates. Compos. Struct. **96**, 833–849 (2013)
52. Mahamood, R.M., Akinlabi, E.T., Shukla, M., Pityana, S.: Evolutionary additive manufacturing: an overview. Lasers Eng. **27**, 161–178 (2014)

MIX
Papier | Fördert gute Waldnutzung
FSC® C083411

Kolibri 360 GmbH
Ferdinand-Jühlke-Straße 7
99095 Erfurt, Deutschland
produktsicherheit@kolibri360.de